*Applied Probability*
*Control*
*Economics*
*Information and Communication*
*Modeling and Identification*
*Numerical Techniques*
*Optimization*

# Applications of Mathematics

# 7

N. N. Vorob'ev



# Game Theory

## Lectures for Economists and Systems Scientists

Translated and supplemented by S. Kotz

Springer-Verlag
New York Heidelberg Berlin

N. N. Vorob'ev
Director of the Mathematical Methods Division
Institute for Social Economical Questions
Academy of Sciences of the USSR
198147 Leningrad
Serpuchovskaya ul. 38, ISEP AN
USSR

S. Kotz
Department of Mathematics
Temple University
Philadelphia, Pennsylvania 19122



AMS Subject Classifications: 90-01, 90D05, 90D10, 90D12

Library of Congress Cataloging in Publication Data

Vorob'ev, Nikolaĭ Nikolaevich, 1925–
Game theory.
(Applications of mathematics ; 7)
Translation of Teoriia igr.
Bibliography: p.
Includes index.
1. Game theory. I. Title.
QA269.V7413 519.3 77–1430
ISBN 0-387-90238-4

The original Russian edition TEORIYA IGR; LEKTSII DLYA EKONOMISTOV-KIBERNETIKOV was published in 1974 by Izvadel'stvo Leningradskogo Universiteta, Leningrad.



Printed in the United States of America.

9 8 7 6 5 4 3 2 1

ISBN 0-387-90238-4 Springer-Verlag New York Heidelberg Berlin
ISBN 3-540-90238-4 Springer-Verlag Berlin Heidelberg New York

# Translator's remark

The translator used lecture notes based on the Russian original of this book during the academic year 1975–1976 at Temple University and found the text suitable for a one semester course for (selected junior and) senior undergraduates or beginning graduate students specializing in applied branches of mathematics or operations research. It was also possible to incorporate within the one semester course some additional material dealing with the actual solution of matrix games. (These topics are covered adequately in many of the elementary texts listed in the Selected Bibliography at the end of this volume.)

For more (mathematically) advanced students, this text served as the core supplemented by about half a dozen research papers. The titles of these research papers may be found in the Basic Survey Papers, Lucas [1] and Vorob'ev [2], [3] given on page 174.

All the exercises are found at the end of the text in compliance with the author's request. This arrangement should not affect students who use this book with their instructor's guidance. Those who use this volume for self-study are urged to refer as often as possible to this part of the text since it contains exercises relevant to almost every section.

S. Kotz

# Preface

The basis for this book is a number of lectures given frequently by the author to third year students of the Department of Economics at Leningrad State University who specialize in economical cybernetics.

The main purpose of this book is to provide the student with a relatively simple and easy-to-understand manual containing the basic mathematical machinery utilized in the theory of games. Practical examples (including those from the field of economics) serve mainly as an interpretation of the mathematical foundations of this theory rather than as indications of their actual or potential applicability.

The present volume is significantly different from other books on the theory of games. The difference is both in the choice of mathematical problems as well as in the nature of the exposition. The realm of the problems is somewhat limited but the author has tried to achieve the greatest possible systematization in his exposition. Whenever possible the author has attempted to provide a game-theoretical argument with the necessary mathematical rigor and reasonable generality.

Formal mathematical prerequisites for this book are quite modest. Only the elementary tools of linear algebra and mathematical analysis are used. The single exception is the fixed-point theorem which is mentioned in Section 3 of Chapter 3; this profound result is, however, quite intuitive and "plausible." The mathematical arguments presented in this volume are not only elementary but with few exceptions, rather simple. However, these cannot be termed as trivial since, when put together, they comprise a somewhat complex logical structure. A clear comprehension of this structure may require some effort on the part of the reader. On the other hand, each separate topic studied in this book can be easily understood; the proofs are given without omitting any logical step in the sequence of the argument and problems are solved in their entirety or at least up to the numerical calculations.

When studying the theory of games, the student is required to absorb a substantial number of new and often unwanted notions. Moreover, the instruction practice of the theory of games for economists and especially a self-study of this subject may be flexible and may allow several different variants and permit selectivity in the topics. For these reasons the author incorporated a certain "concentricity" in the development of this course by repeating later in the book certain notions and even some logical structures introduced in the first chapter, in a more general form.

The amount of material in this volume is more than one can cover in a one hour per week lecture course during a term. The instructor, however, will have no difficulty shifting some of the topics to a laboratory period or to a problem-solving session or can perhaps even exclude some of the material.

Numeration of formulas is carried out separately for each chapter and each formula is noted by a pair of numbers, the first referring to the number of the section in the given chapter, while the second to the consecutive number of the formula in that section.

N.N.V.

# Contents

# Matrix games 1

## 1.1 Definition of a noncooperative game

*1.1.1.* This book discusses primarily the so-called *noncooperative games*. In these games the goal of each participant (player) is to achieve the largest possible individual gain (profit or payoff). Games in which the actions of the players are directed to maximize the gains of "collectives" (coalitions) without subsequent subdivision of the gain among the players within the coalition are called *cooperative* games. The theory of cooperative games in general is much more involved and is not studied in this book.

*1.1.2.* Let $I$ denote the set of all the players. We shall further assume that the set $I$ is finite; however, in the modern theory of games, games with an infinite set of players are studied as well and these are also of practical interest. It is customary to assign a number to each one of the players, i.e. we shall denote the set $I$ as $I=\{1,2,\dots,n\}$.

Let each player $i\in I$ have at his or her disposal a certain set $S_i$ of available actions which are referred to in the theory of games as *strategies*. It is natural to assume that all players have at least two distinct strategies each since if they possess only one strategy their actions are predetermined and they do not actually participate in the game.

The process of the game consists of each one of the players choosing a certain strategy $s_i\in S_i$. Thus, as a result of each "round" of the game, a system of strategies $(s_1,\dots,s_n)=s$ is put together. This system is called a *situation*. The set of all situations is denoted in a natural manner by $S=S_1\times S_2\times\cdots\times S_n$ or equivalently as

$$S=\prod_{i\in I} S_i.$$

In other words the set of all the situations $S$ is the cartesian product of the sets of strategies of all the players.

In each situation $s$ the players attain certain gains (payoffs). The payoff of player $i$ in situation $s$ is usually denoted by $H_i(s)$. The function $H_i$, defined on the set of all the situations, is called the *payoff function* of player $i$.

*1.1.3.* We are now ready to present a precise definition of a noncooperative game.

**Definition.** A system

$$\Gamma = \langle I, \{S_i\}_{i \in I}, \{H_i\}_{i \in I} \rangle, \tag{1.1}$$

where $I$ and $S_i$ ($i \in I$) are sets and $H_i$ are functions defined on the set $S = \prod_{i \in I} S_i$ taking on real values, is called a *noncooperative game*.

*1.1.4.* Among the phenomena described in terms of noncooperative games there are many that lead to a subdivision of a certain fixed (constant) amount of funds among the players. This corresponds in the theory of games to constant-sum games.

**Definition.** A noncooperative game (1.1) is called a *constant-sum game* if there exists a constant $c$ such that $\sum_{i \in I} H_i(s) = c$ for each and every situation $s \in S$.

## 1.2 Admissible situations and the equilibrium situation

*1.2.1.* Intuitively, a situation $s$ in the game (1.1) will be admissible for player $i$ if by replacing his or her present strategy in this situation with some other strategy, player $i$ is unable to increase his or her payoff.

Formally, let $s = (s_1, \ldots, s_{i-1}, s_i, s_{i+1}, \ldots, s_n)$ be an arbitrary situation in game $\Gamma$ and let $s_i$ be a strategy for player $i$. We form a new situation that differs from situation $s$ only in that strategy $s_i$ for player $i$ is now replaced by the strategy $s_i'$. The new situation obtained $(s_1, \ldots, s_{i-1}, s_i', s_{i+1}, \ldots, s_n)$ is commonly denoted by $s \| s_i'$. Obviously, if the strategies $s_i$ and $s_i'$ coincide we have $s \| s_i' = s$.

One can now define an admissible strategy as follows:

**Definition.** A situation $s$ in game (1.1) is called *admissible for the player $i$* if for any other strategy $s_i'$ for this player we have

$$H_i(s \| s_i') \leqslant H_i(s).$$

The term "admissible" can be justified by the fact that if in situation $s$ there exists a strategy $s_i'$ for the player $i$ such that

$$H_i(s \| s_i') > H_i(s),$$

then the player $i$, knowing that the situation $s$ will materialize (for example, he or she may have tentatively arranged with the other players to play in such a way that this situation will indeed take place) may choose strategy $s_i'$ at the last moment and as a result of this choice end up with a larger payoff. In this sense situation $s$ may be viewed as inadmissible for player $i$.

*1.2.2* **Definition.** A situation $s$, which is admissible for all the players is called an *equilibrium situation* (or an *equilibrial situation*).

In other words situation $s$ is an equilibrium situation if the inequality $H_i(s \| s_i') \leqslant H_i(s)$ is satisfied for any player $i$ and for any strategy of this player $s_i' \in S_i$.

It is obvious from the definition that in equilibrium situations and only in such situations no player is interested in deviating from his initial strategy. In particular, if an equilibrium situation results from an agreement between the players then no player is interested in violating such an agreement. On the other hand, if an agreement results in a nonequilibrium situation then—it follows from the definition—at least one player may be interested in breaking a promise.

**Definition.** *An equilibrium strategy* of a player in a noncooperative game is a strategy that appears in at least one equilibrium situation of the game.

A substantial part of the theory of noncooperative games consists of a study of the properties of corresponding equilibrium situations and of equilibrium strategies for players as well as devising methods to determine such situations and strategies.

The process of determination of an equilibrium situation in a noncooperative game is often referred to as the *solution* of a game.

## 1.3 Strategic equivalence of games

*1.3.1.* The great variety of noncooperative games makes it desirable to combine them into classes with the property that all games belonging to one class possess the same basic properties. This will permit us in what follows to concentrate on a particular game of the simplest structure (instead of dealing with all the games in a given class).

Classes of strategically equivalent games are an example of such subdivisions.

**Definition.** Let two noncooperative games with the same sets of players and the same strategies be given (i.e. the games differ only in their payoff functions):

$$\Gamma' = \langle I, \{S_i\}_{i \in I}, \{H_i'\}_{i \in I} \rangle, \tag{3.1}$$

$$\Gamma'' = \langle I, \{S_i\}_{i \in I}, \{H_i''\}_{i \in I} \rangle. \tag{3.2}$$

The games $\Gamma'$ and $\Gamma''$ are called *strategically equivalent* if a positive number $k$ and real numbers $c_i$ exist (for each one of the player $i \in I$) such that for any situation $s$

$$H_i'(s) = kH_i''(s) + c_i. \tag{3.3}$$

The fact that the game $\Gamma'$ is strategically equivalent to the game $\Gamma''$ is denoted by $\Gamma' \sim \Gamma''$.

*1.3.2.* It is easy to verify the following three properties of strategic equivalence.

(1) *Reflexivity*. Each game is strategically equivalent to itself: that is, $\Gamma \sim \Gamma$. To show this it is sufficient to set $k=1$ and $c_i=0$ in (3.3).

(2) *Symmetry*. If $\Gamma \sim \Gamma'$ then $\Gamma' \sim \Gamma$. Indeed it follows from (3.3) that

$$H_i(s) = \frac{1}{k} H_i'(s) - \frac{c_i}{k},$$

where $1/k > 0$.

(3) *Transitivity*. If $\Gamma \sim \Gamma'$ and $\Gamma' \sim \Gamma''$ then $\Gamma \sim \Gamma''$. To prove this assertion we write the conditions of the strategic equivalences for the pairs of games $(\Gamma, \Gamma')$ and $(\Gamma', \Gamma'')$:

$$H_i'(s) = kH_i(s) + c_i, \qquad H_i''(s) = k'H_i'(s) + c_i'.$$

According to the definition, both $k>0$ and $k'>0$. It follows from these inequalities that

$$H_i''(s) = kk'H_i(s) + (k'c_i + c_i'),$$

where, clearly, $k \cdot k' > 0$.

Thus the relation of strategic equivalence is indeed an equivalence relation, namely, it subdivides the totality of all noncooperative games into pairwise disjoint classes of equivalent games.

*1.3.3.* The difference between two strategically equivalent games is basically in the amount of initial capital (funds) $c$ possessed by the players and in the relative units in which the payoffs are measured. (This is indicated by the coefficients $k$.) It is therefore natural to suppose that the rational behavior of players in distinct strategically equivalent games should be the same. In particular, the following theorem is valid.

**Theorem.** *Strategically equivalent games possess the same equilibrium situations.*

PROOF. Let $\Gamma' \sim \Gamma''$, and let $s^*$ be an equilibrium situation in game $\Gamma'$. This means that for all $i \in I$ and $s_i \in S_i$, the inequality

$$H_i'(s^* \| s_i) \leqslant H_i'(s^*).$$

is valid.

Applying (3.3) to the above equality we obtain

$$kH_i''(s^*\|s_i)+c_i \leqslant kH_i''(s^*)+c_i,$$

hence (recalling that $k>0$) we have

$$H_i''(s^*\|s_i) \leqslant H_i''(s^*)$$

for all $i\in I$ and $s_i \in S_i$, which implies that the situation $s^*$ is an equilibrium situation in game $\Gamma''$ defined by (3.2). □

*1.3.4* **Definition.** A noncooperative game

$$\Gamma=\langle I,\{S_i\}_{i\in I},\{H_i\}_{i\in I}\rangle$$

is called a *zero-sum game* if for each situation $s\in S$

$$\sum_{i\in I} H_i(s)=0.$$

**Theorem.** *Any noncooperative constant sum game is strategically equivalent to a certain zero-sum game.*

Proof. Consider the zero-sum game

$$\Gamma'=\langle I,\{S_i\}_{i\in I},\{H_i'\}_{i\in I}\rangle$$

in which the sum of payoffs of all the players is equal to $c$ for every situation of the game and choose arbitrary numbers $c_i$ $(i\in I)$ such that $\Sigma_{i\in I}c_i=c$. Setting $H_i(s)=H_i'(s)-c_i$ we arrive at a zero-sum game. □

## 1.4 Antagonistic games

*1.4.1* **Definition.** The game

$$\Gamma=\langle I,\{S_i\}_{i\in I},\{H_i\}_{i\in I}\rangle$$

is called *antagonistic* if there are only two players in this game and the values of the payoff function for these players in each situation are the same in absolute value but (are) of the opposite sign.

Thus for an antagonistic game we have $I=\{\mathrm{I},\mathrm{II}\}$ and

$$H_{\mathrm{II}}(s)=-H_{\mathrm{I}}(s), \qquad s\in S. \tag{4.1}$$

It follows from (4.1) that in an antagonistic game $H_{\mathrm{I}}(s)+H_{\mathrm{II}}(s)=0$ for any situation $s$. Hence an antagonistic game is also a zero-sum two-person game.

*1.4.2.* Clearly, to define an antagonistic game, it is sufficient to stipulate the payoff function of one of the players only; the payoff function of his or her opponent is determined from (4.1). Thus an antagonistic game is usually denoted by the triple

$$\Gamma=\langle A,B,H\rangle, \tag{4.2}$$

where $A$ and $B$ are the sets of strategies for players I and II, respectively, and $H$ is the payoff function for player I which is a real-valued function defined on the set of pairs of the form $(a,b)$ with $a \in A$ and $b \in B$. It follows from Section 1.3.4 that each noncooperative constant sum two-person game is strategically equivalent to a certain antagonistic game.

## 1.5 Saddle points

*1.5.1.* We shall now redefine the notion of an equilibrium situation for the particular case of an antagonistic game (4.2). Clearly an equilibrium situation in such a game is a situation $(a,b)$ satisfying

$$
\begin{aligned}
&H_{\mathrm{I}}(a,b) \geqslant H_{\mathrm{I}}(a',b), \qquad a' \in A, \\
&H_{\mathrm{II}}(a,b) \geqslant H_{\mathrm{II}}(a,b'), \qquad b' \in B.
\end{aligned}
\tag{5.1}
$$

Taking (4.1) into account we can rewrite the last inequality as

$$-H_{\mathrm{I}}(a,b) \geqslant -H_{\mathrm{I}}(a,b'), \qquad b' \in B,$$

or

$$H_{\mathrm{I}}(a,b) \leqslant H_{\mathrm{I}}(a,b'), \qquad b' \in B.$$

This, together with (5.1), can be written as the double inequality

$$H(a',b) \leqslant H(a,b) \leqslant H(a,b'), \qquad a' \in A, \qquad b' \in B.$$

This inequality expresses the following property of function $H$ at point $(a,b)$: the value of function $H$ can only decrease with the variation of variable $a$ and can only increase with the variation of variable $b$. If we visualize the surface described by function $H$ in coordinates $a$ and $b$, then the saddle points of this surface correspond to the equilibrium situations of the game (cf. Figure 1).

*1.5.2.* It should be noted that the notion of a saddle point in the theory of games differs in two respects from the analogous notion in geometry. First, in geometry the "saddle-shapedness" of a point does not depend on

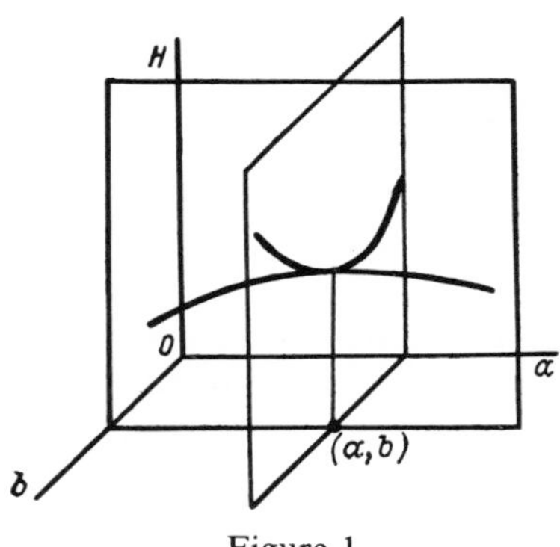


Figure 1

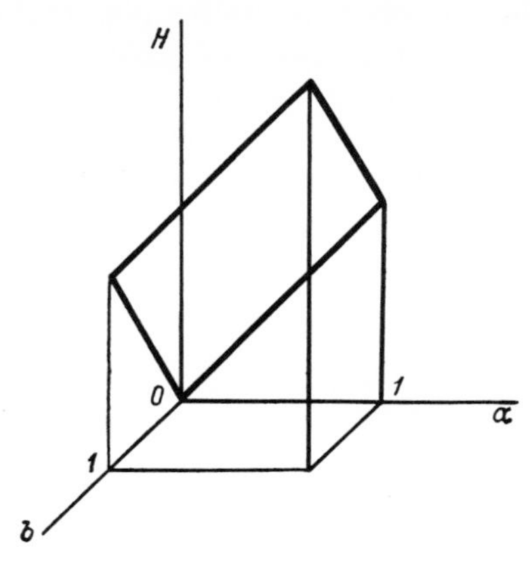


Figure 2

the directions in which a function of two variables increases or decreases. In game theory, however, for a point to be a saddle point it is necessary that it attain the maximum with respect to its *first* coordinate and the minimum with respect to its *second*.

Next, the "saddle-shapedness" of a point in geometry is an analytic property and is associated with the fact that the corresponding derivatives vanish at that point. In game theory this kind of analyticity of the extrema is not required. Moreover, it is often the case that a saddle point turns out to be at the boundary of the domain of the definition of a function. For example, the point $[1,0]$ is a saddle point for the function $H=a+b$ defined on the square $0 \leqslant a, b \leqslant 1$ (cf. Figure 2).

## 1.6 Auxiliary propositions about extrema

*1.6.1* **Definition.** Let a function $f$ be defined on the set $D$. Its *supremum* on this set is the least of the numbers $S$ satisfying $f(x) \leqslant S$ for any $x \in D$. The supremum of a function $f$ on $D$ is denoted by $\sup_{x \in D} f(x)$ or if the argument of the function and the domain of its variation are self-evident, the notation is then simply $\sup f(x)$.

Analogously, an *infimum* of a function $f$ on $D$ is the largest of the numbers $s$ satisfying $s \leqslant f(x)$. The infimum of a function $f$ on $D$ is denoted by $\inf_{x \in D} f(x)$ or if no ambiguity can possibly arise, by $\inf f(x)$.

If the supremum of a function $f$ is actually attained on $D$, i.e. if $x^* \in D$ exists such that $f(x^*) = \sup f(x)$, then it is called the *maximum* and is denoted by $\max f(x)$. If the infimum of a function is attained it is called the *minimum* and is denoted by $\min f(x)$.

*1.6.2.* It follows directly from the definitions above that if a constant $c$ exists such that $f(x) \leqslant c$ for any $x \in D$, then $\sup_{x \in D} f(x) \leqslant c$ also, and if a constant $c$ satisfies $c \leqslant f(x)$ for any $x \in D$, then $c \leqslant \inf_{x \in D} f(x)$ also.

*1.6.3.* If the functions $f$ and $g$ are defined on the same domain $D$ and $f(x) \leqslant g(x)$ for any $x \in D$, then $\sup f(x) \leqslant \sup g(x)$ and $\inf f(x) \leqslant \inf g(x)$.

Indeed it follows from the assumption that $f(x) \leqslant g(x) \leqslant \sup g(x)$ and from Section 1.6.2 we obtain $\sup f(x) \leqslant \sup g(x)$.

The inequality for the infima is proved analogously.

Clearly, if some of the suprema or infima appearing in the statements above are actually attained then, in the assertions proved above, these can be replaced by maxima and minima, respectively.

*1.6.4* **Theorem.** *If $x$ varies in the domain $\mathfrak{X}$ and $y$ varies in the domain $\mathfrak{Y}$, then for any function $f(x,y)$ defined on $\mathfrak{X} \times \mathfrak{Y}$ the following inequality is valid:*

$$\sup_x \inf_y f(x,y) \leqslant \inf_y \sup_x f(x,y). \tag{6.1}$$

Proof. We have for any $x$ and $y$

$$f(x,y) \leqslant \sup_x f(x,y).$$

Consequently, applying the result in Section 1.6.3, we obtain

$$\inf_y f(x,y) \leqslant \inf_y \sup_x f(x,y).$$

The right-hand side of the last inequality is a constant. Hence in view of Section 1.6.2,

$$\sup_x \inf_y f(x,y) \leqslant \inf_y \sup_x f(x,y). \qquad \square$$

**Corollary.** *If the outer extrema are attained in* (6.1), *then*

$$\max_x \inf_y f(x,y) \leqslant \min_y \sup_x f(x,y). \tag{6.2}$$

*If, moreover, all the inner extrema are attained, i.e. if for any $x$ there exists $\min_y f(x,y)$ and for any $y$ there exists $\max_x f(x,y)$, then*

$$\max_x \min_y f(x,y) \leqslant \min_y \max_x f(x,y). \tag{6.3}$$

Inequalities (6.1), (6.2), and (6.3) are sometimes called the *minimax inequalities*. They play an important role in the theory of games.

## 1.7 Minimax equalities and saddle points

*1.7.1* **Theorem.** *In order that the function $f(x,y)$ defined on the product $\mathfrak{X} \times \mathfrak{Y}$ possess saddle points, it is necessary and sufficient that the following minimaxes exist (i.e. are attained)*

$$\max_x \inf_y f(x,y), \qquad \min_y \sup_x f(x,y), \tag{7.1}$$

*and that the equality*

$$\max_x \inf_y f(x,y) = \min_y \sup_x f(x,y) \tag{7.2}$$

*be satisfied.*

PROOF (*Necessity*). Let the function $f$ possess saddle points and let $(x^*,y^*)$ be one of those points. This implies that

$$f(x,y^*) \leqslant f(x^*,y^*) \leqslant f(x^*,y). \tag{7.3}$$

The expression $f(x^*,y^*)$ is a constant. Applying the assertion proved in Section 1.6.2 to the left-hand side of the above inequality, we have

$$\sup_x f(x,y^*) \leqslant f(x^*,y^*) \tag{7.4}$$

and furthermore

$$\inf_y \sup_x f(x,y) \leqslant \sup_x f(x,y^*) \leqslant f(x^*,y^*). \tag{7.5}$$

Applying the same argument to the right-hand side of inequality (7.3) we finally obtain

$$\inf_y \sup_x f(x,y) \leqslant \sup_x f(x,y^*) \leqslant f(x^*,y^*) \leqslant \inf_y f(x^*,y) \leqslant \sup_x \inf_y f(x,y), \tag{7.6}$$

i.e.

$$\inf_y \sup_x f(x,y) \leqslant \sup_x \inf_y f(x,y).$$

However, in view of the theorem in Section 1.6.4 the reverse inequality is also valid. Therefore

$$\inf_y \sup_x f(x,y) = \sup_x \inf_y f(x,y). \tag{7.7}$$

Hence in inequality (7.6) the extermal terms are equal. Consequently, all the parts of this inequality are equal, in particular,

$$\inf_y \sup_x f(x,y) = \sup_x f(x,y^*).$$

Thus the infimum is attained in the expression $\inf_y \sup_x f(x,y)$ (at $y = y^*$) and this expression can be written as $\min_y \sup_x f(x,y)$. In the same manner it follows from (7.6) that

$$\inf_y f(x^*,y) = \sup_x \inf_y f(x,y), \tag{7.8}$$

and thus one can replace $\sup_x \inf_y f(x,y)$ by $\max_x \inf_y f(x,y)$.

In view of the above equality, (7.7) can be written now as

$$\max_x \inf_y f(x,y) = \min_y \sup_x f(x,y).$$

*Sufficiency*. Now let the minimaxes (7.1) exist and be equal and let the outer extrema in these expressions be attained at the points $x^*$ and $y^*$, respectively. This means that

$$\max_x \inf_y f(x,y) = \inf_y f(x^*,y).$$

Moreover, we also have $\inf_y f(x^*,y) \leqslant f(x^*,y^*)$ so that

$$\max_x \inf_y f(x,y) = \inf_y f(x^*,y) \leqslant f(x^*,y^*) \tag{7.9}$$

and analogously

$$f(x^*,y^*) \leqslant \sup_x f(x,y^*) = \min_y \sup_x f(x,y). \tag{7.10}$$

In view of the assumption that the minimaxes (7.1) are equal, all the terms in expressions (7.9) and (7.10) are equal. In particular, $\sup_x f(x,y^*) = f(x^*,y^*)$. This means that we have for any $x$

$$f(x,y^*) \leqslant f(x^*,y^*). \tag{7.11}$$

Similarly, it follows from (7.9) that $\inf_y f(x^*,y) = f(x^*,y^*)$, whence

$$f(x^*,y^*) \leqslant f(x^*,y). \tag{7.12}$$

Inequalities (7.11) and (7.12) imply that $(x^*,y^*)$ is a saddle point for the function $f$. □

*1.7.2 Remark.* In the course of the proof of the theorem it was established that as the components of the saddle point, one can choose $x$ and $y$ —independently—at which the outer extrema are attained in the minimaxes (7.1). Therefore, if $(x_1,y_1)$ and $(x_2,y_2)$ are two saddle points of function $f$, so are the points $(x_1,y_2)$ and $(x_2,y_1)$.

This property of the set of all the saddle points of a function is usually referred to as a *rectangular* property. Clearly, every rectangular subset of the Cartesian product $\mathfrak{X} \times \mathfrak{Y}$ is of the form $\mathfrak{X}^* \times \mathfrak{Y}^*$, where $\mathfrak{X}^* \subset \mathfrak{X}$ and $\mathfrak{Y}^* \subset \mathfrak{Y}$.

Next it follows from (7.6) and (7.7) that the value of the function at a saddle point is equal to the common value of the minimaxes (7.1). Consequently, the values of a function at each one of its saddle points are the same.

## 1.8 Matrix games

*1.8.1* **Definition.** Antagonistic games in which each player possesses a finite number of strategies are called *matrix games*.

This name is due to the fact that such games can be described as follows: Consider a rectangular array in which the rows correspond to the strategies for the first player and the columns to the strategies for the second, and the cells in the array located at the intersection of the rows and columns correspond to the situations of the game. If we place in each cell the payoff of the *first* player in the appropriate situation, we then obtain the description of the game in the form of a certain matrix. This matrix is called the *matrix of the game* or the *payoff matrix*.

The actual process of "playing a matrix game" is conveniently represented as follows: Let matrix **A** be given; player I chooses a certain row of this matrix, and player II selects a certain column. These choices are made by the players independently of one another. After the choices are made, player I gets the payoff equal to the number appearing at the intersection of the selected row and column. Clearly, if this number is negative, this signifies that player I loses a certain amount.

A matrix game is completely determined by its payoff matrix. Therefore we shall sometimes refer to a "game with the payoff matrix **A**" simply as "game **A**."

*1.8.2.* The following notation and terminology is commonly used. Let

$$\mathbf{A} = \begin{Vmatrix} a_{11} & a_{12} & \cdots & a_{1n} \\ a_{21} & a_{22} & \cdots & a_{2n} \\ \vdots & & & \\ a_{m1} & a_{m2} & \cdots & a_{mn} \end{Vmatrix}$$

be a matrix of a certain game. In this case the game is called the $m \times n$ *game*. The strategies of the first player are labelled by the ordinal numbers of the corresponding rows and the strategies of player II by the (ordinal) numbers of the columns. The $i$th row of matrix **A** is denoted by $A_{i\cdot}$ and its $j$th column by $A_{\cdot j}$.

Clearly, the pair $(i,j)$, where $i$ is the row number and $j$ the column number, constitute a situation in a matrix game.

*1.8.3.* A situation $(i^*, j^*)$ in a matrix game is an equilibrium situation (since a matrix game is antagonistic, such a situation can be referred to as a saddle point) if for any $i = 1, \ldots, m$ and for any $j = 1, \ldots, n$

$$a_{ij^*} \leqslant a_{i^*j^*} \leqslant a_{i^*j}.$$

*1.8.4.* In view of the discussion presented in Section 1.6, in order that a matrix game possess a saddle point it is necessary and sufficient that the following two minimaxes be equal:

$$\max_i \min_j a_{ij} = \min_j \max_i a_{ij}.$$

(Since the sets of strategies of each one of the players are finite in our case, the minimaxes exist trivially.)

*1.8.5.* The determination of saddle points of matrix **A** can be carried out using the following schematic procedure:

$$\begin{array}{cccccc}
\left.\begin{array}{cccc} a_{11} & a_{12} & \cdots & a_{1n} \\ a_{21} & a_{22} & \cdots & a_{2n} \\ \vdots & a_{m2} & \cdots & a_{mn} \end{array}\right] &
\begin{array}{l} \to \min_j a_{1j} \\ \to \min_j a_{2j} \\ \to \min_j a_{mj} \end{array} & \Bigg\} \; \max_i \min_j a_{ij} \\
\begin{array}{cccc} \downarrow & \downarrow & & \downarrow \\ \max_i a_{i1} & \max_i a_{i2} & \cdots & \max_i a_{in} \end{array} & & \\
\underbrace{\hphantom{\max_i a_{i1} \quad \max_i a_{i2} \quad \cdots \quad \max_i a_{in}}}_{\min_j \max_i a_{ij}} & &
\end{array}$$

## 1.9 Mixed strategies

*1.9.1.* If the minimaxes of the elements of a matrix (i.e. the quantities $\max_i \min_j a_{ij}$ and $\min_j \max_i a_{ij}$) are not equal to each other, then in view of the theorem proved in Section 1.6.4 the game with such a matrix does *not* possess an equilibrium situation.

If this is the case, player I can assure his payoff to be $\max_i \min_j a_{ij}$, while player II plays in such a manner that player I receives no more than $\min_j \max_i a_{ij}$. The problem of how the difference $\min_j \max_i a_{ij} - \max_i \min_j a_{ij}$ should be subdivided between the players (this difference is positive in the case under consideration) thus remains open. It is therefore natural that the players should seek in these cases additional strategic opportunities in order to assure for themselves the largest possible share of this difference. It turns out that it is desirable that they choose their strategies for this purpose randomly.

*1.9.2* **Definition.** A random variable whose values are the strategies of a player is called a *mixed strategy* for the player.

Thus the determination of a mixed strategy for a player consists of assigning probabilities with which his or her original strategies are selected.

If a player chooses one strategy with probability 1 and any one of the remaining strategies with probability 0, then he or she has chosen in fact one definite strategy. Hence each one of the initial strategies for a player is a fortiori a mixed strategy as well. To single out these strategies in the class of all mixed strategies, they are often referred to as *pure* strategies.

Since a mixed strategy for a player is determined by the probabilities of that player choosing the available pure strategies, it can be represented as the vector

$$X = (x_1, x_2, \ldots, x_m), \tag{9.1}$$

where

$$x_1 \geqslant 0, \qquad x_2 \geqslant 0, \ldots, x_m \geqslant 0, \tag{9.2}$$

$$\sum_{i=1}^{m} x_i = 1. \tag{9.3}$$

*1.9.3.* The last condition can be written more compactly. Denote by $J_n$ the $n$-dimensional vector with all the components equal to 1:

$$J_n = (1, \ldots, 1).$$

Then using scalar products, equality (9.3) can be rewritten as

$$XJ_n^T = 1,$$

where $T$ is the transposition operation (in this case, it transposes a row vector into a column vector).

*1.9.4.* As it is known, the totality of all the vectors of type (9.1) forms an $n$-dimensional Euclidean space. The set of these vectors subject to conditions (9.2) and (9.3) form an $(n-1)$-dimensional simplex spanned by the unit vectors

$$E^{(1)} = (1, 0, \ldots, 0),$$
$$E^{(2)} = (0, 1, \ldots, 0),$$
$$\vdots$$
$$E^{(n)} = (0, 0, \ldots, 1).$$

We refer to this simplex as the *fundamental simplex* and denote it by $S_n$.

For $n=2$ the simplex is a segment (Figure 3); for $n=3$ it is a rectangle (Figure 4); and for $n=4$ it becomes a tetrahedron.

In some cases it is convenient to consider the simplex $S_n$ on its own, separately from the over-all Euclidean space (of which this simplex is a part). In this case the coordinates $x_1, \ldots, x_n$ of the *points* on the simplex can be viewed as nonnegative masses that should be allocated at the vertices of the simplex in such a manner that the center of gravity of these masses is located at a *given* point. Therefore the coordinates $x_1, \ldots, x_n$ of the points

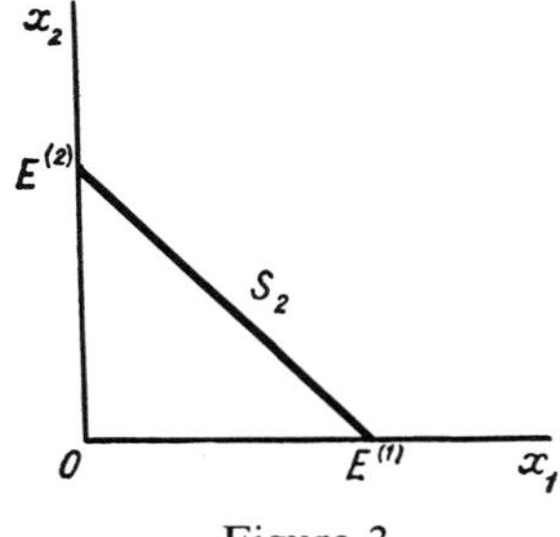


Figure 3

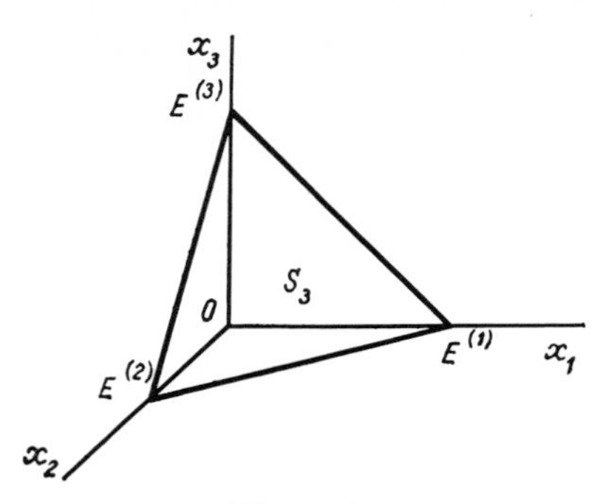


Figure 4

on a simplex are often called the *barycentric coordinates*. In our case the masses correspond to the probabilities with which the pure strategies are selected.

Pure strategies are identified by the property that they correspond to points with one of the barycentric coordinates being unity, while all the others are zero. Clearly, these points are the vertices of the fundamental simplex.

## 1.10 A mixed extension of a game

*1.10.1.* Let players I and II choose their mixed strategies

$$X=(x_1,\ldots,x_m),\qquad Y=(y_1,\ldots,y_n)$$

independently of one another in a game with the payoff matrix

$$\mathbf{A}=\begin{bmatrix} a_{11} & a_{12} & \cdots & a_{1n} \\ a_{21} & a_{22} & \cdots & a_{2n} \\ \vdots & & & \\ a_{m1} & a_{m2} & \cdots & a_{mn} \end{bmatrix}.$$

**Definition.** A pair $(X, Y)$ of mixed strategies for the players in a matrix game is called a *situation in mixed strategies*.

In a situation with mixed strategies $(X, Y)$, each usual situation $(i,j)$ (in pure strategies) becomes a random event occurring with probability $x_i y_j$. Since in the situation $(i,j)$ player I receives payoff $a_{ij}$, the mathematical expectation of his or her payoff under $(X, Y)$ is equal to

$$\sum_{i=1}^{m}\sum_{j=1}^{n} a_{ij}x_i y_i. \tag{10.1}$$

This number is assumed to be the payoff of player I in a situation in mixed strategies $(X, Y)$ and is denoted by $H(X, Y)$.

We thus arrive at a new game which can be described as follows:

**Definition.** *A mixed extension* of a matrix game is called the antagonistic game $\langle S_m, S_n, H\rangle$ in which the set of strategies for the players is the set of their mixed strategies in the original game and the payoff function for player I is defined by expression (10.1).

Utilizing scalar products one can rewrite expression (10.1) in the form

$$\sum_{i=1}^{m} x_i \sum_{j=1}^{n} a_{ij} y_j = \sum_{i=1}^{m} x_i A_{i\cdot} Y^T = X\mathbf{A}Y^T.$$

*1.10.2.* Recalling the definition of a saddle point in an antagonistic game, we observe that the situation $(X^*, Y^*)$ in a mixed extension of a matrix game will be a saddle point (i.e. an equilibrium situation) provided for any $X \in S_m$ and $Y \in S_n$ the following double inequality is satisfied:

$$X\mathbf{A}Y^{*T} \leqslant X^*\mathbf{A}Y^{*T} \leqslant X^*\mathbf{A}Y^T. \tag{10.2}$$

*1.10.3* **Lemma** (on the transition to mixed strategies). *If $Y$ is an arbitrary strategy for player* II *and $a$ is a number such that*

$$A_{i\cdot}Y^T \leqslant a, \qquad i=1,\ldots,m, \tag{10.3}$$

*then for any mixed strategy $X=(x_1,\ldots,x_m)$ of player* I

$$X\mathbf{A}Y^T \leqslant a.$$

Proof. Multiplying termwise both sides of each one of the inequalities (10.3) by $x_i$ (for $i=1,\ldots,m$), we obtain

$$x_i A_{i\cdot} Y^T \leqslant x_i a$$

(since $x_i \geqslant 0$ the same inequality sign is preserved). Adding up these inequalities we have

$$\sum_{i=1}^{m} x_i A_{i\cdot} Y^T = X\mathbf{A}Y^T \leqslant a \sum_{i=1}^{m} x_i = a. \qquad \square$$

*Remark.* Transitions to mixed strategies in the inequalities of the form

$$\begin{aligned} A_{i\cdot}Y^T &\geqslant a, \qquad i=1,\ldots,m,\\ XA_{\cdot j} &\leqslant a, \qquad j=1,\ldots,n,\\ XA_{\cdot j} &\geqslant a, \qquad j=1,\ldots,n \end{aligned} \tag{10.3'}$$

are carried out analogously.

*1.10.4* **Theorem.** *In order that the situation $(X^*, Y^*)$ be an equilibrium situation it is necessary and sufficient that for all $i=1,\ldots,m$ and $j=1,\ldots,n$*

*the equality*

$$A_{i\cdot}Y^{*T} \leqslant X^*\mathbf{A}Y^{*T} \leqslant X^*A_{\cdot j} \tag{10.4}$$

*be satisfied.*

PROOF. The necessity is self-evident since inequality (19.4) is a particular case of inequality (10.2).

To prove sufficiency we apply to both sides of inequality (10.4) the lemma on transition to mixed strategies (Section 1.10.3). This immediately yields inequality (10.2). □

*1.10.5* **Theorem.** *If situation $(i^*,j^*)$ is an equilibrium situation for the matrix game* **A**, *it is also an equilibrium situation for its mixed extension.*

PROOF. The equilibrium property of a situation $(i^*,j^*)$ in the matrix game **A** implies that for any $i=1,\ldots,m$ and $j=1,\ldots,n$

$$a_{ij^*} \leqslant a_{i^*j^*} \leqslant a_{i^*j}. \tag{10.5}$$

Now (10.4) becomes (10.5) if we substitute the corresponding pure strategies $i^*$ and $j^*$ into (10.4) in place of $X^*$ and $Y^*$.

*1.10.6.* It turns out that for any matrix game there exists an equilibrium situation in a *mixed* extension of a game.

In view of the discussion in Section 1.6, in order to prove this assertion it is sufficient to verify the existence and the equality of the corresponding minimaxes

$$\max_X \inf_Y X\mathbf{A}Y^T \quad \text{and} \quad \min_Y \sup_X X\mathbf{A}Y^T.$$

Actually we shall prove below the existence and equality of the minimaxes

$$\max_X \min_Y X\mathbf{A}Y^T \quad \text{and} \quad \min_Y \max_X X\mathbf{A}Y^T.$$

## 1.11 Existence of minimaxes in mixed strategies

*1.11.1* **Lemma.** *For any $Y_0 \in S_n$ there exists the maximum* $\max_X X\mathbf{A}Y_0^T$ *and for any $X_0 \in S_m$ there exists the minimum* $\min_Y X_0\mathbf{A}Y^T$.

PROOF. We have

$$X\mathbf{A}Y_0^T = \sum_i x_i A_{i\cdot} Y_0^T.$$

This implies that $X\mathbf{A}Y_0^T$ is a linear function in variables $x_1,\ldots,x_m$ and hence a continuous function in these variables. Therefore it attains its maximum on the closed and bounded (i.e. compact) set $S_m$. □

The second part of the lemma is proved analogously.

*1.11.2* **Lemma.** *For any $X_0 \in S_m$ there exists $j_0$ (depending on $X_0$) such that*

$$\min_Y X_0\mathbf{A}Y^T = X_0 A_{\cdot j_0}.$$

*For any* $Y_0 \in S_n$ *there exists* $i_0$ *such that*

$$\max_X X\mathbf{A}Y_0^T = A_{i_0\cdot}Y_0.$$

PROOF. Consider the numbers $X_0A_{\cdot j}$ $(j=1,\ldots,n)$ and let $X_0A_{\cdot j_0}$ be the smallest of these numbers. This implies that

$$X_0A_{\cdot j_0} \leqslant X_0A_{\cdot j}, \qquad j=1,\ldots,n.$$

Utilizing the lemma on transition to mixed strategies (Section 1.10.3) we obtain

$$X_0A_{\cdot j_0} \leqslant X_0\mathbf{A}Y^T.$$

The last inequality is valid for any $Y \in S_n$. Therefore in view of the discussion given in Section 1.6.2 we have

$$X_0A_{\cdot j_0} \leqslant \min_Y X_0\mathbf{A}Y^T. \tag{11.2}$$

On the other hand, $X_0A_{\cdot j_0}$ is a number of the form $X_0\mathbf{A}Y^T$ (namely, when $Y$ is the pure strategy $j_0$). Hence

$$X_0A_{\cdot j_0} \geqslant \min_Y X_0\mathbf{A}Y^T. \tag{11.3}$$

Combining inequalities (11.2) and (11.3) we arrive at (11.1). □

*1.11.3* **Lemma.** *The quantity*

$$\max_X X\mathbf{A}Y^T \tag{11.4}$$

*is a continuous function of* $Y$ *and the quantity* $\min_Y X\mathbf{A}Y^T$ *is a continuous function of* $X$.

PROOF. We shall prove only the first assertion of the lemma. As indicated in the preceding item, it is sufficient to prove that

$$\max_i A_{i\cdot}Y^T, \tag{11.5}$$

viewed as a function $Y$, is a continuous function in this variable.

Clearly, (as was mentioned above) the scalar product $A_{i\cdot}Y^T$ for any $i$ is everywhere continuous in $Y$. Since $i$ takes on only a finite number of values, all the scalar products $A_{i\cdot}Y^T$ are equi-continuous in $Y$.

Take an arbitrary $\varepsilon > 0$ and choose $\delta$ such that for $|Y' - Y''| < \delta$,

$$|A_{i\cdot}Y'^T - A_{i\cdot}Y''^T| < \varepsilon \tag{11.6}$$

or $A_{i\cdot}Y'^T - \varepsilon < A_{i\cdot}Y''^T < A_{i\cdot}Y'^T + \varepsilon$. In accordance with 1.6.3 therefore $\max_i A_{i\cdot}Y'^T - \varepsilon < \max_i A_{i\cdot}Y''^T < \max_i A_{i\cdot}Y'^T + \varepsilon$ so that

$$\left|\max_i A_{i\cdot}Y'^T - \max_i A_{i\cdot}Y''^T\right| < \varepsilon,$$

which proves the continuity of (11.5). □

*1.11.4* **Theorem.** *The minimaxes*

$$\max_X \min_Y X\mathbf{A}Y^T \quad \text{and} \quad \min_Y \max_X X\mathbf{A}Y^T$$

*exist.*

PROOF. As it was shown above $\max_X X\mathbf{A}Y^T$ is a continuous function of $Y$. This function, defined on a finite dimensional closed bounded set $S_n$, achieves its minimum in the set; i.e. the minimax $\min_Y \max_X X\mathbf{A}Y^T$ is attained.

An analogous argument shows that the maximin $\max_X \min_Y X\mathbf{A}Y^T$ is also attained. □

## 1.12 Convex sets

*1.12.1* **Definition.** A subset $S$ of a vector space is called a *convex set* (in this vector space) if for any $U$ and $V \in S$ and for an arbitrary $\lambda \in [0,1]$ the vector $\lambda U + (1-\lambda)V \in S$.

Convex sets play an important role in the theory of games. The following properties of convex sets will be utilized in what follows.

*1.12.2.* If $S$ is a convex set, $X_1, X_2, \ldots, X_k \in S$ and $\alpha_1, \alpha_2, \ldots, \alpha_k$ are nonnegative numbers satisfying $\sum_{i=1}^k \alpha_i = 1$, then the vector $\sum_{i=1}^k \alpha_i X_i$, called a convex *combination* of vectors $X_1, X_2, \ldots, X_n$, also belongs to $S$. The totality of all convex combinations of a given set of vectors is called the *convex hull* of these vectors.

*1.12.3.* If $S$ and $T$ are two disjoint convex sets, then there exists a hyperplane which subdivides them, i.e. there exists a hyperplane $VZ = c$ such that

$$\begin{aligned} VZ &\geqslant c \quad \text{for } Z \in S. \\ VZ &< c \quad \text{for } Z \in T. \end{aligned}$$

(here $V$ is the "directional vector" of the plane).

*1.12.4* **Definition.** A point $U$ is called an *extreme point* of a convex set $S$ if $U \in S$ and there are no two distinct points $U'$ and $U''$ belonging to $S$ such that $U = \frac{1}{2}(U' + U'')$.

If $S$ is a convex bounded nonempty set possessing a finite number of extreme points, then each point of this set can be represented as a convex combination of extreme points.

*1.12.5.* A convex cone is a particular case of a convex set.

**Definition.** *A convex cone* in a vector space is a set $C$ such that for any $U'$ and $U'' \in C$ and any nonnegative numbers $\lambda'$ and $\lambda''$

$$\lambda' U' + \lambda'' U'' \in C.$$

Clearly the null-vector $\mathbf{0}$ always belongs to an arbitrary convex cone.

**Definition.** Let the vectors $U_1, U_2, \ldots, U_n$ be given; then the convex cone of all the vectors of the form

$$\sum_{i=1}^{n} \lambda_i U_i, \quad \text{where } \lambda_i \geqslant 0 \text{ for } i=1,\ldots,n,$$

is called the cone *spanned by the vectors* $U_1, \ldots, U_n$.

## 1.13 The lemma on two alternatives

*1.13.1* **Lemma** (two alternatives). *Given an arbitrary matrix* $\mathbf{A}$ *one of the following two possibilities (i.e. alternatives) holds:*

(1) *a vector* $X \in S_m$ *exists such that* $XA_{.j} \geqslant 0$ *for all* $j=1,\ldots,n$;

(2) *a vector* $Y \in S_n$ *exists such that* $A_{i.}Y^T \leqslant 0$ *for all* $i=1,\ldots,m$.

PROOF. Consider the convex hull of the fundamental simplex $S_m$ (i.e. of all the vectors $E^{(1)}, \ldots, E^{(m)}$) and the vectors $A_{.j}$. Denote this hull by $C$. Two possibilities may present themselves:

(a) $\mathbf{0} \notin C$ (Figure 5). In this case, in view of the argument in Section 1.12.3, the point $\mathbf{0}$ can be separated from the set $C$ by a certain hyperplane. We may assume that this hyperplane passes through the point $\mathbf{0}$ and that the whole set $C$ is located on one side of this plane. Let $Vz=0$ be the equation of this hyperplane and for any $z \in C$ we have (without loss of generality)

$$Vz > 0. \tag{13.1}$$

In particular, $VE^{(i)} > 0$ for all $i=1,\ldots,m$. Consider the numbers $VE^{(i)} = v_i$ $(i=1,\ldots,m)$. Since all these numbers are positive, so is their sum:

$$v = \sum_{i=1}^{m} v_i > 0.$$

Now consider the vector

$$X = \left( \frac{v_1}{v}, \ldots, \frac{v_m}{v} \right).$$

Clearly $X \in S_m$.

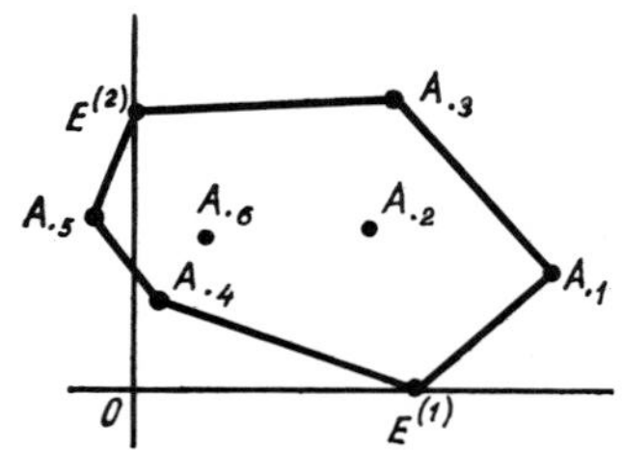


Figure 5

Since $v > 0$, it follows from (13.1) that

$$Xz = \frac{1}{v} Vz > 0$$

for any point $z \in S$. In particular, this is true for all the points $A_{\cdot j}$; thus $XA_{\cdot j} > 0$ $(j = 1, \ldots, n)$ and the required vector $X$ is constructed.

(b) $\mathbf{0} \in C$ (Figure 6). In this case the point $\mathbf{0}$ can be represented as a convex combination of the vertices of the polyhedron $C$, i.e. as a convex combination of the points $A_{\cdot 1}, \ldots, A_{\cdot n}, E^{(1)}, \ldots, E^{(m)}$. Let

$$\sum_{j=1}^{n} \alpha_j A_{\cdot j} + \sum_{i=1}^{m} \eta_i E^{(i)} = \mathbf{0}. \tag{13.2}$$

Here the quantities $\alpha_1, \ldots, \alpha_n, \eta_i, \ldots, \eta_m$ are all $\geqslant 0$ and

$$\sum_{j=1}^{n} \alpha_j + \sum_{i=1}^{m} \eta_i = 1. \tag{13.3}$$

The equality (13.2) is presented in vector form. Coordinatewise this equality states:

$$\sum_{j=1}^{n} \alpha_j a_{ij} + \eta_i = 0. \tag{13.4}$$

Since $\eta_i \geqslant 0$ we have

$$\sum_{j=1}^{n} \alpha_j a_{ij} \leqslant 0. \tag{13.5}$$

Next it follows that

$$\alpha = \sum_{j=1}^{n} \alpha_j \geqslant 0. \tag{13.6}$$

If the sum $\alpha = \sum_{j=1}^{n} \alpha_j = 0$ were equal to 0, it would have to follow from the fact that all $\alpha_j$ is nonnegative, that each one of these numbers is zero: $\alpha_j = 0$ $(j = 1, \ldots, n)$. This conclusion, in view of (13.4), would imply that all $\eta_j$ $(j = 1, \ldots, n)$ are zero, which contradicts (13.3). Thus (13.6) indeed holds with a strict inequality sign.

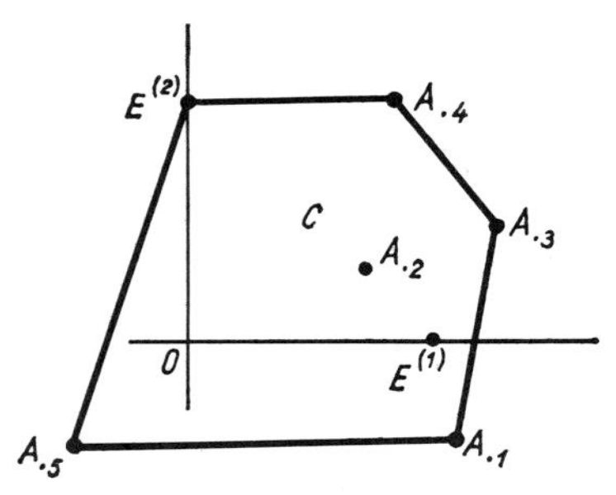


Figure 6

Now set $\alpha_j/\alpha = y_j$ and construct the vector $Y=(y_1,\ldots,y_n)$. It is easily seen that $Y \in S_n$.

Since $\alpha > 0$ we can divide the left-hand side of (13.5) by this number without violating the inequality sign. As a result of this division we get

$$\sum_{j=1}^{n} y_j a_{ij} = A_{i\cdot} Y^T \leqslant 0 \quad \text{for all } i=1,\ldots,m.$$

Thus $Y$ is the required vector and the lemma is proved. □

## 1.14 The minimax theorem

*1.14.1* **Theorem.** *For any matrix* **A** *the equality*

$$\max_X \min_Y X\mathbf{A}Y^T = \min_Y \max_X X\mathbf{A}Y^T$$

*is valid.*

Proof. We shall apply to matrix **A** the lemma on two alternatives proved in Section 1.13.

Assume that the first alternative holds, i.e. a vector $X_0$ exists such that

$$X_0 A_{\cdot j} \geqslant 0, \qquad j=1,\ldots,n. \tag{14.1}$$

Passing on to mixed strategies we have, in view of Section 1.10.3,

$$X_0 \mathbf{A} Y^T \geqslant 0 \qquad (Y \in S_n).$$

Since this inequality is valid for any vector $Y \in S_n$, we have in accordance with Section 1.6.2

$$\min_Y X_0 \mathbf{A} Y^T \geqslant 0 \tag{14.2}$$

and a fortiori $\max_X \min_Y X\mathbf{A}Y^T \geqslant 0$.

Now let the second alternative of the lemma be valid, i.e. let a vector $Y_0$ exist such that

$$A_{i\cdot} Y_0^T \leqslant 0, \qquad i=1,\ldots,m. \tag{14.3}$$

Passing on to mixed strategies we obtain $X\mathbf{A}Y_0^T \leqslant 0$ for all $X \in S_m$. Therefore

$$\max_X X\mathbf{A}Y_0^T \leqslant 0$$

also and a fortiori

$$\min_Y \max_X X\mathbf{A}Y^T \leqslant 0. \tag{14.4}$$

We thus observe that depending on whether the first or the second alternative is valid either inequality (14.2) or inequality (14.4) holds. Hence at least one of these inequalities is satisfied. It is therefore impossible that both inequalities are violated. In other words, the double inequality

$$\max_X \min_Y XAY^T < 0 < \min_Y \max_X XAY^T \tag{14.5}$$

*cannot* hold.

Now choose an arbitrary real number $t$ and consider the matrix

$$\mathbf{A}(t) = \begin{bmatrix} a_{11}-t & a_{12}-t & \dots & a_{1n}-t \\ a_{21}-t & a_{22}-t & \dots & a_{2n}-t \\ \vdots & & & \\ a_{m1}-t & a_{m2}-t & \dots & a_{mn}-t \end{bmatrix}.$$

For the matrix $\mathbf{A}(t)$ we rewrite inequality (14.5) as

$$\max_X \min_Y X\mathbf{A}(t)Y^T < 0 < \min_Y \max_X X\mathbf{A}(t)Y^T. \tag{14.6}$$

However,

$$X\mathbf{A}(t)Y^T = \sum_{i=1}^{m} \sum_{j=1}^{n} x_i(a_{ij}-t)y_j$$

$$= \sum_{i=1}^{m} \sum_{j=1}^{n} x_i a_{ij} y_i - t \sum_{i=1}^{m} \sum_{j=1}^{n} x_i y_j = X\mathbf{A}Y^T - t.$$

Therefore (14.6) is equivalent to

$$\max_X \min_Y (X\mathbf{A}Y^T - t) < 0 < \min_Y \max_X (X\mathbf{A}Y^T - t)$$

or

$$\max_X \left( \min_Y X\mathbf{A}Y^T - t \right) < 0 < \min_Y \left( \max_X X\mathbf{A}Y^T - t \right)$$

or

$$\max_X \min_Y X\mathbf{A}Y^T - t < 0 < \min_Y \max_X X\mathbf{A}Y^T - t$$

or, finally,

$$\max_X \min_Y X\mathbf{A}Y^T < t < \min_Y \max_X X\mathbf{A}Y^T.$$

From the above argument, both sides of the last inequality cannot be valid simultaneously no matter what the number $t$ is. This implies that there exists no number between the two minimaxes $\max_X \min_Y X\mathbf{A}Y^T$ and $\min_Y \max_X X\mathbf{A}Y^T$ that is strictly larger than the first one and strictly smaller than the second. This implies that the second minimax cannot exceed the first one, i.e.

$$\min_Y \max_X X\mathbf{A}Y^T \leqslant \max_X \min_Y X\mathbf{A}Y^T. \tag{14.7}$$

However, in view of the argument presented in Section 1.6 the reverse inequality is always satisfied:

$$\max_X \min_Y X\mathbf{A}Y^T \leqslant \min_Y \max_X X\mathbf{A}Y^T. \tag{14.8}$$

Combining (14.7) and (14.8) we obtain

$$\max_X \min_Y X\mathbf{A}Y^T = \min_Y \max_X X\mathbf{A}Y^T,$$

which proves the theorem. □

## 1.15 The value of the game and optimal strategies

*1.15.1.* Player I, participating in a game with matrix **A**, may argue as follows: "Assume that I am going to choose strategy $i$, then in the worst case my payoff will be $\min_j a_{ij}$. Therefore it would be reasonable for me to choose a strategy $i$ such that this minimum is maximized: $\max_i \min_j a_{ij}$. This maxmin I am definitely bound to get even in the least favorable case when I know nothing about my opponent's intentions."

Similarly, player II may argue as follows: "Assume that I am going to choose strategy $j$, then in the worst situation I am going to lose the amount $\max_i a_{ij}$. It would therefore seem reasonable for me to choose strategy $j$ such that this maximum is minimized: $\min_j \max_i a_{ij}$. Acting in this manner, I would not allow my opponent to win more than the value of this minimax even in the least favorable case when my opponent knows everything about my intentions."

These arguments explain inter alia from the "common sense approach" the inequality

$$\max_i \min_j a_{ij} \leqslant \min_j \max_i a_{ij}$$

as well as the equality of these minimaxes, which is valid when the saddle point exists (cf. Section 1.8.4).

If we now pass on to a mixed extension of a game, the above argument should be somewhat modified. Player 1 now argues as follows: "Assume I am going to select strategy $X$. Then in the worst case I shall receive the amount $\min_Y X\mathbf{A}Y^T$. Caution compels me to choose a strategy such that the maximum

$$\max_X \min_Y X\mathbf{A}Y^T \tag{15.1}$$

is achieved. And I shall definitely obtain this maximum no matter what the circumstances are."

An analogous argument on behalf of player II will lead him or her to the "reasonable" conclusion that he or she (can and) *should not* pay player I an amount that exceeds

$$\min_Y \max_X X\mathbf{A}Y^T. \tag{15.2}$$

But as was shown in Section 1.14, the maximin (15.1) and minimax (15.2) are equal. Consequently, if each one of the players behaves reasonably ("rationally"), the payoff of the second player to the first will be a definite amount equalling each one of the minimaxes (15.1) and (15.2).

*1.15.2.* Thus if one permits the usage of mixed strategies, then the outcome of a matrix game turns out to be essentially predetermined: it does not depend on skill or on the depth of psychological reasoning carried out by the players, but depends solely on the conditions of the game that are completely determined by assigning the matrix **A**. This is the reason that matrix games are referred to as *completely determined games* (or games with *complete information*).

We thus see that a certain number that expresses the natural payoff of the first player to the second corresponds to any matrix. This correspondence can be viewed as a function defined on the set of all matrix games (i.e. on the set of all the matrices) and the value of this function, for each particular matrix, is the common value of the minimaxes (15.1) and (15.2).

**Definition.** The common value of minimaxes (15.1) and (15.2) is called the *value* of the matrix game with the payoff matrix **A** and is denoted by $v(\mathbf{A})$.

*1.15.3.* The choice of strategy $X$ for player I at which the outer maximum is attained in (15.1) represents the best available action for that player. Similarly, the choice of strategy for player II at which the outer minimum is attained in (15.2) is the most advisable one. This, however, means that players I and II should select as their strategies those which form a pair representing a saddle point.

**Definition.** Equilibrium strategies of players in an antagonistic game are called their *optimal* strategies.

A *solution* of a matrix game is the process of determining the value of the game and the pairs of optimal strategies. On the other hand, the saddle points of an antagonistic game are also called the solutions of a game. Clearly, this dual usage of the term "solution" is justified and cannot cause ambiguity or misunderstanding.

*1.15.4.* We can now rewrite the characterizing inequality for a saddle point $(X^*, Y^*)$ in the following form:

$$X\mathbf{A}Y^{*T} \leqslant v \leqslant X^*\mathbf{A}Y^T \quad \text{for } X \in S_m \text{ and } Y \in S_n.$$

We observe that if player I chooses his or her optimal strategy, he or she gets a payoff not less than the value of the game, no matter what action player II undertakes. Analogously, by choosing his or her optimal strategy, player II assures that his or her loss will always be less than the value of the game. Consequently, it would not make sense for one of the players to conceal his or her *optimal* strategy from the other.

*1.15.5.* An important remark is in order: The "determinicity" of a game mentioned above should be understood in the sense that when mixed strategies are utilized, the *mathematical expectation* of the payoff due to player I is uniquely determined. Clearly, the actual payoffs to the players in the individual "rounds" of a game will be in general different from this expected payoff.

## 1.16 Three properties of the value of a game

*1.16.1* **Theorem.** *In a matrix game with the payoff matrix* **A** *the value of the game* $v(\mathbf{A})$ *satisfies*

$$v(\mathbf{A}) = \max_X \min_j XA_{\cdot j} = \min_Y \max_i A_{i\cdot} Y^T, \tag{16.1}$$

*and the outer extrema are attained at the optimal strategies of the players.*

PROOF. We have

$$v(\mathbf{A}) = \max_X \min_Y X\mathbf{A}Y^T.$$

In view of Section 1.11.2, for each $X$ there exists $j_0$ such that for all $j = 1, \ldots, n$

$$\min_Y X\mathbf{A}Y^T = XA_{\cdot j_0} \leqslant XA_{\cdot j}.$$

Consequently,

$$\min_Y X\mathbf{A}Y^T = \min_j XA_{\cdot j}.$$

Since this equality is valid for any $X \in S_m$, one must have

$$\max_X \min_Y X\mathbf{A}Y^T = \max_X \min_j XA_{\cdot j},$$

and the outer extrema on the left- and right-hand sides are attained for the *same* value of $X$. It is known, however, that on the left-hand side the outer extremum is attained at the optimal strategies of player I. Therefore the same strategies yield the extremum on the right-hand side. This proves the first half of the lemma. The second part [i.e. the equality $v(\mathbf{A}) = \min_Y \max_i A_{i\cdot} Y^T$] is proved using an analogous symmetric argument. □

*1.16.2* **Theorem.** *For any matrix* **A** *we have*

$$\max_i \min_j a_{ij} \leqslant v(\mathbf{A}) \leqslant \min_j \max_i a_{ij}. \tag{16.2}$$

PROOF. It follows from (16.1) that

$$v(\mathbf{A}) = \max_X \min_j XA_{\cdot j} \geqslant \max_i \min_j a_{ij}$$

because maximization with respect to the set of pure strategies instead of the set of all mixed strategies reduces the domain of maximization and thus can only possibly decrease the outer maximum. This proves the left-hand side of inequality (16.2). The right-hand side of this inequality is proved in an analogous manner. □

*1.16.3* **Theorem.** *If player* I *possesses a pure optimal strategy* $i_0$, *then*

$$v = \max_i \min_j a_{ij} = \min_j a_{i_0 j}. \tag{16.3}$$

*If player* II *possesses a pure optimal strategy* $j_0$, *then*

$$v = \min_j \max_i a_{ij} = \max_i a_{ij_0}. \tag{16.3$'$}$$

To verify (16.3) it is sufficient to recall that in view of the first theorem (Section 1.16.1) the maximum in the case under consideration is attained for the pure strategy $X = i_0$. This implies that

$$v(\mathbf{A}) = \max_i \min_j a_{ij} = \min_j a_{i_0 j}. \qquad \square$$

The game-theoretical meaning of this theorem is quite evident. Player I gets the maximin payoff if his or her opponent knows the pure strategy used by player I. However, if a certain pure strategy of player I is optimal, which means that he or she will confine playing action to this strategy and thus his or her opponent plays, taking this fact into consideration. Loosely speaking, one can assert that the presence of a pure optimal strategy for a player reflects certain unfavorable conditions of the game—for this player—since this does not permit "confusing" his or her opponent.

## 1.17 An example: 2×2 games

*1.17.1.* Consider a matrix game with only two pure strategies for each one of the players. The payoff matrix $\mathbf{A}$ corresponding to this game is of the form

$$\mathbf{A} = \begin{pmatrix} a_{11} & a_{12} \\ a_{21} & a_{22} \end{pmatrix}.$$

Let $X$ be an arbitrary mixed strategy for player I (in particular this strategy may be a pure strategy). If $x$ is the probability that player I chooses his first pure strategy in the mixed strategy $X$, then this player chooses his second pure strategy with probability $1-x$. Therefore the strategy $X$ can be represented as $(x, 1-x)$. Analogously, if $Y$ is an arbitrary mixed strategy of player II it is of the form $(y, 1-y)$. Thus strategy $X$ is uniquely determined by the number $x$ and strategy $Y$ by the number $y$. Pure strategies correspond to the values 0 or 1 of the "parameters" $x$ and $y$. In view of the above we shall denote the situation $(X, Y)$ by the pair $(x,y)$.

We thus have

$$XAY^T=(x,1-x)\begin{pmatrix} a_{11} & a_{12} \\ a_{21} & a_{22}\end{pmatrix}\begin{pmatrix} y \\ 1-y\end{pmatrix}$$
$$=xya_{11}+x(1-y)a_{12}+(1-x)ya_{21}+(1-x)(1-y)a_{22}$$
$$=xy(a_{11}-a_{12}-a_{21}+a_{22})+x(a_{12}-a_{22})+y(a_{21}-a_{22})+a_{22}.$$

Let $(x^*,y^*)$ be a saddle point of the game under consideration. This implies that

$$\left.\begin{matrix} H(0,y^*) \\ H(1,y^*)\end{matrix}\right\} \leqslant H(x^*,y^*) \leqslant \left\{\begin{matrix} H(x^*,0), \\ H(x^*,1).\end{matrix}\right. \tag{17.1}$$

*1.17.2.* At first let us assume that player I possesses a pure optimal strategy. For definiteness, let the optimal pure strategy be the first one (otherwise we change the numbering of the strategies), i.e. let $x^*=1$. In this case inequality (17.1) becomes

$$H(0,y^*) \leqslant H(1,y^*) \leqslant \left\{\begin{matrix} H(1,0), \\ H(1,1),\end{matrix}\right.$$

or substituting the explicit expression for the payoff function we get

$$a_{21}y^*+a_{22}(1-y^*) \leqslant a_{11}y^*+a_{12}(1-y^*) \leqslant \left\{\begin{matrix} a_{12}, \\ a_{11}.\end{matrix}\right. \tag{17.2}$$

If $a_{12}>a_{11}$, the right-hand side of this inequality becomes

$$a_{12}+y^*(a_{11}-a_{12}) \leqslant a_{11};$$

this inequality can be valid only if $y^*=1$. Hence in this case player II possesses a unique optimal strategy that is his or her first pure strategy. The value of the game is equal to $a_{11}$.

Similarly, if $a_{12}<a_{11}$ the right-hand side of (17.2) becomes

$$a_{12}+y^*(a_{11}-a_{12}) \leqslant a_{12},$$

which implies that $y^*=0$. In this case the unique optimal strategy for player II is his or her second pure strategy and the value of the game is equal to $a_{12}$.

Finally, if $a_{12}=a_{11}$, then the right-hand side of (17.2) becomes $a_{11} \leqslant a_{11}$, which is an identity and so no restrictions of the value of $y^*$ are imposed. Consequently, in this case an optimal strategy of player II is given by any $y^*$ provided this number satisfies the left-hand side of inequality (17.2):

$$y^*(a_{21}+a_{12}-a_{11}-a_{22}) \leqslant a_{12}-a_{21}. \tag{17.3}$$

Since, as we already know, player II must have an optimal strategy, inequality (17.3) has at least one solution on the closed interval [0, 1]. Moreover, since inequality (17.3) restricts $y^*$ only from one side, the set of solutions of this inequality is an interval in [0, 1] containing at least one of its end points, i.e. point 0 or point 1 or possibly both of these points. The value of the game is $a_{11}=a_{12}$.

Incidentally, the following feature has been confirmed: If in a $2\times 2$ game, player I possesses a pure optimal strategy, then so does player II. From symmetry considerations, if in a $2\times 2$ game player II possesses a pure optimal strategy, so does player I.

*1.17.3.* Now let player I possess no pure optimal strategy in a $2\times 2$ game. In view of the remark at the end of Section 1.17.2, this implies that player II also has no pure optimal strategy.

However, it follows from the existence theorem (Section 1.15) that both players do possess optimal strategies. Hence in this case the optimal strategies are mixed.

Let $x^*$ and $y^*$ be arbitrary mixed optimal strategies of players I and II, respectively. Since these strategies are not pure, we have

$$0<x^*, \qquad y^*<1 \tag{17.4}$$

We again write the equilibrium condition for the situation $(x^*,y^*)$:

$$H(x,y^*)\leqslant H(x^*,y^*)\leqslant H(x^*,y).$$

This implies that

$$\max_x H(x,y^*)=H(x^*,y^*)=\min_y (x^*,y).$$

In view of (17.4) the maximum and the minimum in the last formula should be attained at an interior point of the interval [0, 1], namely, these extrema are "analytic." Hence the corresponding partial derivatives should vanish at the points $x^*$ and $y^*$:

$$\left.\frac{\partial H(x,y^*)}{\partial x}\right|_{x=x^*}=0, \qquad \left.\frac{\partial H(x^*,y)}{\partial y}\right|_{y=y^*}=0. \tag{17.5}$$

However,

$$\left.\frac{\partial H(x,y^*)}{\partial x}\right|_{x=x^*}=y^*(a_{11}-a_{12}-a_{21}+a_{22})+a_{12}-a_{22}$$

and

$$\left.\frac{\partial H(x^*,y)}{\partial y}\right|_{y=y^*}=x^*(a_{11}-a_{12}-a_{21}+a_{22})+a_{21}-a_{22}.$$

The equalities (17.5) can thus be rewritten as

$$\begin{aligned} &y^*(a_{11}-a_{12}-a_{21}+a_{22})+a_{12}-a_{22}=0,\\ &x^*(a_{11}-a_{12}-a_{21}+a_{22})+a_{21}-a_{22}=0. \end{aligned} \tag{17.6}$$

In view of the existence theorem, each of these equations has a solution in $x^*$ and $y^*$ that are located in the interval [0, 1]. Moreover, if $a_{11}-a_{12}-a_{21}+a_{22}\neq 0$, each of the equations (17.6) possesses a unique solution. We now show that in the case under consideration the equality

$$a_{11}-a_{12}-a_{21}+a_{22}=0 \tag{17.7}$$

cannot possibly hold.

Indeed (17.7) implies

$$a_{11}+a_{22}=a_{21}+a_{12}. \tag{17.8}$$

For definiteness let us assume that

$$a_{11} \geqslant a_{12}. \tag{17.9}$$

In this case, in view of (17.8)

$$a_{21} \geqslant a_{22}. \tag{17.10}$$

Now if $a_{22} \leqslant a_{12}$ we have, on account of (17.9),

$$a_{22} \leqslant a_{12} \leqslant a_{11},$$

namely, $(x^*,y^*)=(1,0)$ is a saddle point in pure strategies, which contradicts the assumptions stipulated in this case.

If, on the other hand, $a_{22} \geqslant a_{12}$, then (17.10) implies that

$$a_{12} \leqslant a_{22} \leqslant a_{21},$$

and we again arrive at an equilibrium situation in pure strategies [specifically the situation $(x^*,y^*)=(0,0)$], which is also a contradiction. This shows that (17.7) cannot possibly be valid and the optimal strategies of the players are unique.

Thus, since (17.7) is invalid, the system (17.6) possesses the unique solution

$$x^*=\frac{a_{22}-a_{12}}{a_{11}-a_{12}-a_{21}+a_{22}}, \tag{17.11}$$

$$y^*=\frac{a_{22}-a_{21}}{a_{11}-a_{12}-a_{21}+a_{22}}. \tag{17.12}$$

We now evaluate the value of the game:

$$\begin{aligned} v(\mathbf{A}) &= H(x^*,y^*) \\ &= x^*y^*(a_{11}-a_{12}-a_{21}+a_{22})+x^*(a_{12}-a_{22})+y^*(a_{21}-a_{22})+a_{22}, \end{aligned}$$

Replacing $x^*$ and $y^*$ by their explicit expressions given by (17.11) and (17.12), we obtain after some elementary algebraic manipulations

$$v(\mathbf{A})=\frac{a_{11}a_{22}-a_{12}a_{21}}{a_{11}-a_{12}-a_{21}+a_{22}}. \tag{17.13}$$

*1.17.4.* We have thus shown that a solution for a 2×2 game can be carried out using the following procedure:

(1) First check whether an equilibrium situation exists in pure strategies. For this purpose compute and compare the minimaxes

$$\max_i \min_j a_{ij} \quad \text{with} \quad \min_j \max_i a_{ij}. \tag{17.14}$$

If these are equal, the equilibrium situation in pure strategies does exist and the resolution of the game is carried out as indicated above in Section 1.17.1.

(2) If the minimaxes (17.14) are not equal, the game possesses a unique equilibrium situation in the mixed strategies and the optimal strategies of the players as well as the value of the game are directly computed using formulas (17.11), (17.12), and (17.13).

## 1.18 A graphical solution of $2\times n$ games

*1.18.1.* Consider a game in which player I possesses two pure strategies while player II possesses an arbitrary number $n$ of pure strategies. The payoff matrix is now of the form

$$\mathbf{A}=\begin{pmatrix} a_{11} & a_{12} & \dots & a_{1n} \\ a_{21} & a_{22} & \dots & a_{2n} \end{pmatrix}. \tag{18.1}$$

The fundamental simplex of mixed strategies of player I in this case is the closed interval $[0,1]$ in which the coordinate of a point representing a mixed strategy is the probability of using the first pure strategy.

Let player II select his pure strategy $j$. Then the payoff of player I will depend on his or her chosen mixed strategy $X$, i.e. on the probability $x$ of choosing the first pure strategy:

$$XA_{\cdot j}=xa_{1j}+(1-x)a_{2j}.$$

On a graph this relationship between the payoff and $x$ is expressed by a straight line. A particular straight line (Figure 7) corresponds to each pure strategy $j$ of player II. Clearly, if the matrix (18.1) has columns that are numerically the same, then the straight lines corresponding to these columns, i.e. strategies for player II, will coincide. For simplicity we consider all the columns with the same components as a single strategy.

The graph of the function

$$z=\min_j XA_{\cdot j}=\min_j\left(xa_{1j}+(1-x)a_{2j}\right)$$

is the lower envelope of all the straight lines corresponding to the strategies

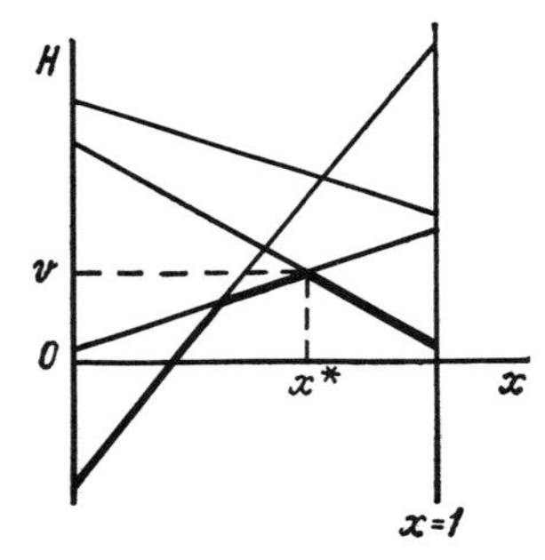


Figure 7

for player II (this envelope is marked by a boldface line in Figure 7). Clearly, such a graph is a broken line that is convex upwards. An upper peak of this broken line corresponds to the value at which the maximum

$$\max_x \min_j XA_{\cdot j} = \max_x \min_j (xa_{1j} + (1-x)a_{2j})$$

is attained.

Thus the abscissa of this point is the optimal mixed strategy for player I and the ordinate of this point represents the value of the game $v(\mathbf{A})$. If there are several of such upper peaks, then the envelope will contain a horizontal segment (cf. Figure 8). The set of optimal strategies for player I will then consist of abscissas of all these points.

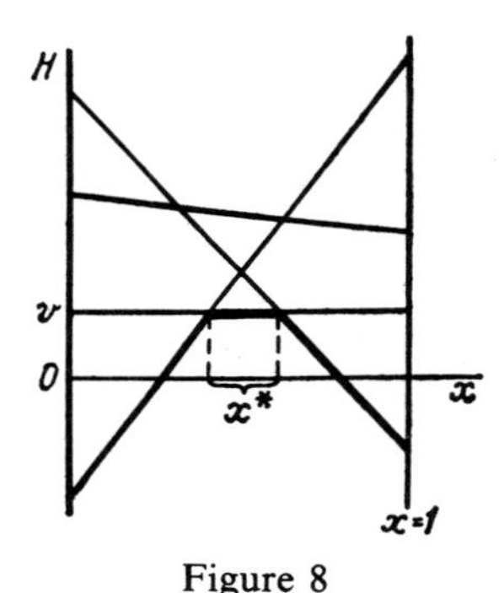


Figure 8

*1.18.2.* The above construction also gives a method for determining optimal strategies for player II. Here several distinct cases are encountered.

First, let the enveloping line possess an upper *horizontal* segment corresponding to the pure strategy $j_0$ of player II. Obviously, this can happen only if $a_{1j_0} = a_{2j_0}$. In this case player II possesses a unique optimal strategy that is a pure strategy.

Assume now that the enveloping line ends up (or begins) with a peak. If the abscissa of the peak is either 0 or 1 (Figure 9), then the optimal strategy for player I is a pure strategy (in Figure 9 the peak is at the point 0) and the optimal strategies of player II are those pure strategies that correspond to the straight lines approaching the peak with a positive slope. Clearly, all the mixtures of these pure strategies will also be strategies for player II.

An analogous situation is valid if the peak point has abscissa 1.

Finally, assume that the abscissa of the peak differs from either 0 or 1. This implies that at least two lines intersect at this peak, one with a positive slope and the other with a negative one (Figure 10). Let

$$z = a_{2j_2} + x(a_{1j_2} - a_{2j_2})$$

$$z = a_{2j_1} + x(a_{1j_1} - a_{2j_1}),$$

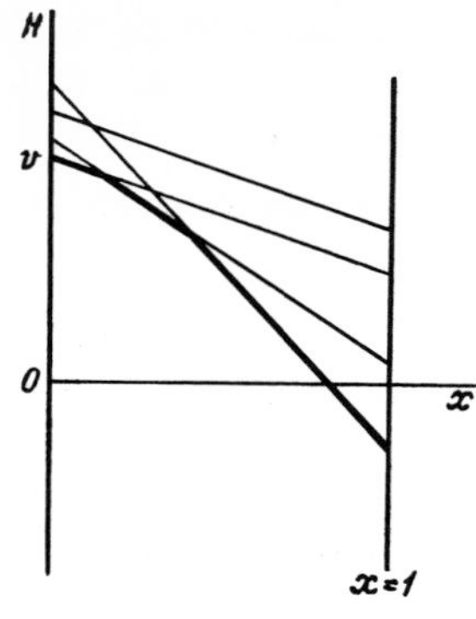


Figure 9

be these two lines. If player II ignores all his other strategies (except strategies $j_1$ and $j_2$), then in the $2\times 2$ game obtained, the value of the game and the unique optimal strategy for player I will be as in the original $2\times m$ game. This implies that by utilizing only the two strategies $j_1$ and $j_2$, player II can prevent player I from obtaining a payoff larger than the value $v(\mathbf{A})$. Consequently, an optimal strategy for player II in the original game can be obtained as a mixture of his two pure strategies $j_1$ and $j_2$. Thus an optimal strategy for player II in the "new" $2\times 2$ game is also his optimal strategy in the original $2\times n$ game. To compute this strategy one can utilize formula (17.12), which in this case becomes

$$y=\frac{a_{2j_2}-a_{2j_1}}{a_{1j_1}-a_{1j_2}-a_{2j_1}+a_{2j_2}}.$$

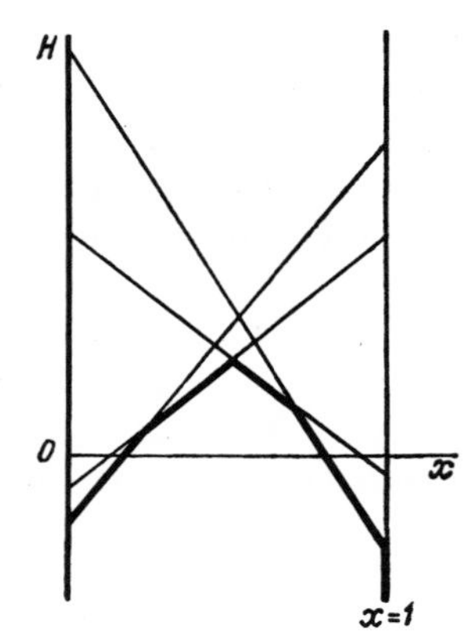


Figure 10

## 1.19 A graphical solution of $m \times 2$ games

*1.19.1.* Now let player II possess two pure strategies and player I have an arbitrary number $m$ of pure strategies. In this case the payoff matrix becomes

$$\mathbf{A} = \begin{bmatrix} a_{11} & a_{12} \\ a_{21} & a_{22} \\ \vdots & \\ a_{m1} & a_{m2} \end{bmatrix}.$$

The resolution of a game of this type is similar to the procedure described in the previous section:

An arbitrary mixed strategy for player II is of the form $Y = (y, 1-y)$, and the totality of these strategies is described by the interval $[0, 1]$.

If player I chooses his or her $i$th pure strategy and player II his or her mixed strategy $Y$, then the payoff to player I is clearly equal to

$$A_{i\cdot}Y^T = a_{i1}y + a_{i2}(1-y).$$

Thus the dependence of this payoff on $y$ is represented graphically by a straight line.

The graph of

$$\max_i A_{i\cdot}Y^T = \max_i (a_{i1}y + a_{i2}(1-y))$$

will be the upper envelope of all straight lines corresponding to the pure strategies of player I (Figure 11). The abscissa of the lowest point on this broken line is the value $y^*$ corresponding to the optimal strategy of player II, while the ordinate represents the value of the game $v(\mathbf{A})$.

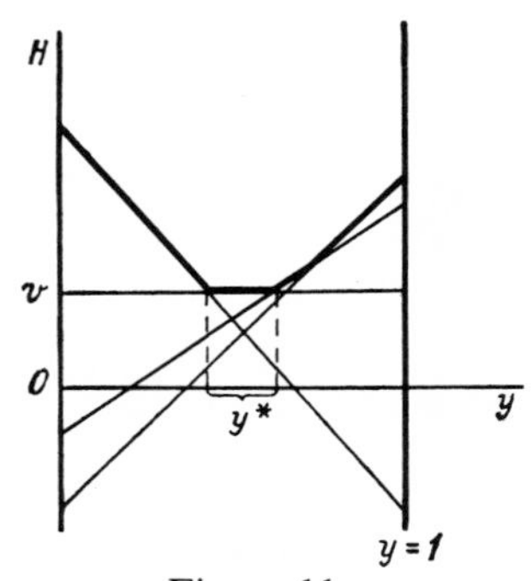


Figure 11

*1.19.2.* A graphical procedure similar to the one described above can be utilized to solve games in which one of the players possesses three pure strategies. However, the construction in this case is rather cumbersome and methods of descriptive geometry are utilized. (Details are given in Ventzel [7E].)

If each of the players possesses more than three pure strategies, a graphical solution of the game is practically impossible.

## 1.20 Sufficient criteria for the value of a game and optimal strategies

*1.20.1.* The following theorems are often useful for checking the optimality of strategies.

**Theorem.** *Let $v(\mathbf{A})$ be the value of a game and let $v$ be an arbitrary number.*
*Let $X_0$ be a strategy for player* I, *then*

$$v \leqslant X_0 A_{\cdot j}, \qquad j=1,\ldots,n, \tag{20.1}$$

*implies that $v \leqslant v(A)$.*
*If $Y_0$ is a strategy for player* II, *then*

$$A_{i\cdot} Y_0^T \leqslant v, \qquad i=1,\ldots,m,$$

*implies that $v(\mathbf{A}) \leqslant v$.*

We shall prove the first part of the theorem. Let $Y^*$ be an optimal strategy for player II. Applying the lemma on the transition to mixed strategies (Section 1.10.3) to (20.1), we have

$$v \leqslant X_0 \mathbf{A} Y^{*T}.$$

However, $X_0 \mathbf{A} Y^{*T} \leqslant v(\mathbf{A})$, yielding the required result. □

*1.20.2* **Theorem.** *Let $X_0$ and $Y_0$ be strategies for players* I *and* II, *correspondingly, and let $v$ be a number such that for any $i=1,\ldots,m$ and $j=1,\ldots,n$,*

$$A_{i\cdot} Y_0^T \leqslant v \leqslant X_0 A_{\cdot j}, \tag{20.2}$$

*then $v$ is the value of the game and $(X_0, Y_0)$ is a saddle point.*

Proof. Applying the lemma on transition to mixed strategies to both sides of inequality (20.2), we obtain

$$X_0 \mathbf{A} Y_0^T \leqslant v \leqslant X_0 \mathbf{A} Y_0^T,$$

i.e. $v = X_0 \mathbf{A} Y^T$.
Substituting this expression for $v$ into (20.2), we have

$$A_{i\cdot} Y_0^T \leqslant X_0 \mathbf{A} Y_0^T \leqslant X_0 A_{\cdot j}$$

for all $i$ and $j$, i.e. in view of the theorem in Section 1.10.4, $(X_0, Y_0)$ is a saddle point and $v$ is the value of the game. □

*1.20.3* **Theorem.** *If for two strategies $X_0$ and $Y_0$ for players* I *and* II, *respectively, the inequality*

$$\max_i A_{i\cdot} Y_0^T \leqslant \min_j X_0 A_{\cdot j} \tag{20.3}$$

*is valid, then $(X_0, Y_0)$ is a saddle point and* (20.3) *holds with an equality sign.*

PROOF. From the definitions of a maximum and a minimum and in view of (20.3), we have

$$A_{i\cdot}Y_0^T \leqslant \max_i A_{i\cdot}Y_0^T \leqslant \min_j X_0 A_{\cdot j} \leqslant X_0 A_{\cdot j}$$

for all $i=1,\dots,m$ and $j=1,\dots,n$. Therefore, in view of the preceding theorem (Section 1.20.2), $(X_0, Y_0)$ is a saddle point and each side of the inequality (20.3) represents the value of the game. Hence these sides are equal to each other. □

*1.20.4* **Theorem.** *If $X_0$ and $Y_0$ are strategies for players* I *and* II, *respectively, and if for any $i=1,\dots,m$ and $j=1,\dots,n$*

$$A_{i\cdot}Y_0^T \leqslant X_0 A_{\cdot j}, \tag{20.4}$$

*then $(X_0, Y_0)$ is a saddle point.*

PROOF. Since inequality (20.4) is satisfied for all $i$ and $j$ we have

$$\max_i A_{i\cdot}Y_0^T \leqslant \min_j X_0 A_{\cdot j};$$

this in turn brings us to the assumptions of the preceding theorem [cf. (20.3)]. □

*1.20.5* **Theorem.** *Let $v$ be the value of a game. If for a strategy $X_0$ for player* I,

$$v \leqslant X_0 A_{\cdot j}, \qquad j=1,\dots,n, \tag{20.5}$$

*then $X_0$ is an optimal strategy for player* I.

*If for a strategy $Y_0$ for player* II,

$$A_{i\cdot}Y_0^T \leqslant v, \qquad i=1,\dots,m,$$

*then $Y_0$ is an optimal strategy for this player.*

PROOF. Let $Y^*$ be an optimal strategy for player II. It then follows from (20.5) and the definition of an optimal strategy that

$$A_{i\cdot}Y^{*T} \leqslant v \leqslant X_0 A_{\cdot j},$$

and the theorem in Section 1.20.2 now yields the optimality of $X_0$.

The second part of the theorem is proved analogously. □

*1.20.6* **Theorem.** *Let $v$ be the value of a game. If for a strategy $X_0$ for player* I

$$v \leqslant \min_j X_0 A_{\cdot j}, \tag{20.6}$$

*then $X_0$ is an optimal strategy for player* I.

*If for a strategy $Y_0$ for player* II

$$\max A_{i\cdot}Y_0^T \leqslant v,$$

*then $Y_0$ is an optimal strategy for this player.*

Proof. It follows from (20.6) that $v \leqslant X_0 A_{\cdot j}$ for all $i = 1, \ldots, n$. Applying the preceding theorem, we obtain that $X_0$ is an optimal strategy.

The second part of the theorem is proved analogously. □

*1.20.7* **Corollary.** *If $v$ is the value of a game, then equality*

$$v = \min_j X_0 A_{\cdot j}$$

*is a necessary and sufficient condition for optimality of the strategy $X_0$ for player* I.

*Analogously*

$$v = \max_i A_{i\cdot} Y_0^T$$

*is a necessary and sufficient condition for optimality of the strategy $Y_0$ for player* II.

Proof. Necessity of the condition follows from the theorem in 1.16.1, while sufficiency is implied by the preceding theorem.

*1.20.8* **Theorem.** *Let $v$ be the value of a game.*

*If $X^* = (x_1^*, \ldots, x_m^*)$ is an arbitrary optimal strategy for player* I *and if for some $j_0$*

$$X^* A_{\cdot j_0} > v, \tag{20.7}$$

*then for any optimal strategy $Y^* = (y_1^*, \ldots, y_n^*)$ for player* II *we have $y_{j_0}^* = 0$.*

*If $Y^* = (y_1^*, \ldots, y_n^*)$ is an arbitrary optimal strategy for player* II *and if for some $i_0$*

$$A_{i_0 \cdot} Y^{*T} < v,$$

*then for any optimal strategy $X^* = (x_1^*, \ldots, x_m^*)$ for player* I *we have $x_{i_0}^* = 0$.*

We shall prove the first part of the theorem. We have

$$X^* A_{\cdot j} \geqslant v \quad \text{for all } j = 1, \ldots, n. \tag{20.8}$$

Now multiply each one of the inequalities (20.8) (*except* the one corresponding to $j = j_0$) by $y_j^*$ and add them up. We thus obtain

$$\sum_{j \neq j_0} X^* A_{\cdot j} y_j^* \geqslant v \sum_{j \neq j_0} y_j^*. \tag{20.9}$$

Now let $j = j_0$. Assuming that $y_{j_0}^* \neq 0$, (20.7) yields

$$X^* A_{\cdot j_0} y_{j_0}^* > v y_{j_0}^*$$

in this case. Adding this inequality termwise to inequality (20.9), we obtain

$$\sum_{j=1}^{n} X^* A_{\cdot j} y_j^* = X^* \mathbf{A} Y^{*T} = v > v \sum_{j=1}^{n} y_j^* = v,$$

which is a contradiction. □

*1.20.9.* The proof of the converse for the above theorem is much more involved. First we shall prove the following lemma known in the literature as *Farkas's lemma* [7P].

**Lemma.** *Let $Z^1, \ldots, Z^n$ and $Z$ be $m$-vectors such that for an arbitrary $m$-vector $W$, the inequality $WZ^{i^T} \geqslant 0$ for all $i=1,\ldots,n$ implies*

$$WZ \geqslant 0.$$

*Then $Z$ belongs to the convex cone $C$ generated by the vectors $Z^1, \ldots, Z^n$.*

PROOF. Assume that $Z \notin C$. This implies that there exist a vector $W$ and a number $d$ such that

$$WU^T > d \quad \text{for all } U \in C \tag{20.10}$$

and $WZ^T = d$.

Clearly $\mathbf{0} \in C$. Therefore (20.10) implies that $d < 0$.

Suppose now that $S_0 \in C$ exists such that $WS_0^T < 0$. If we then multiply $S_0$ by a sufficiently large scalar $t$ (the product $tS_0$ also belongs to the cone $C$), we can assure that $W(tS_0)^T$ would be less than any fixed negative number and, in particular, less than $d$, i.e.

$$W(tS_0)^T < d.$$

This contradicts the condition (20.10) which is valid for all $U \in C$. (Thus the assumption that $Z \notin C$ leads to a contradiction and the lemma is proved.) □

*1.20.10.* We are now ready to prove the converse statement.

**Theorem.** *Let $v$ be the value of a game. If for any optimal strategy $Y^* = (y_1^*, \ldots, y_n^*)$ for player* II *$y_{j_0}^* = 0$, then an optimal strategy $X^* = (x_1^*, \ldots, x_m^*)$ exists for player* I *such that*

$$X^* A_{\cdot j_0} > v. \tag{20.11}$$

The *proof* is quite similar to that of the lemma on two alternatives (cf. Section 1.13.1). Without loss of generality we may assume that $v = 0$.

Consider the cone $C$ spanned by the following $n+m-1$ vectors: the unit vectors $E^{(i)}$ and the columns of matrix $A_{\cdot j}$ for $j \neq j_0$.

First, let us assume that $-A_{\cdot j_0} \in C$. Then the vector $-A_{\cdot j_0}$ can be represented as a linear combination of the vectors that span the cone $C$:

$$-A_{\cdot j_0} = \sum_{i=1}^{n} \eta_i E^{(i)} + \sum_{j \neq j_0} \alpha_j A_{\cdot j}, \tag{20.12}$$

here $\eta_i \geqslant 0$ $(i=1,\ldots,n)$, $\alpha_j \geqslant 0$ $(j \neq j_0)$. Coordinatewise, the equality (20.12) becomes

$$-a_{ij_0} = \eta_i + \sum_{j \neq j_0} \alpha_j a_{ij} \quad \text{for } i = 1, \ldots, m;$$

whence

$$\sum_{j \neq j_0} \alpha_j a_{ij} + a_{ij_0} = -\eta_i \leqslant 0 \quad \text{for } i = 1, \ldots, m.$$

Dividing the last inequality termwise by $1 + \Sigma_{j \neq j_0} \alpha_j > 0$ and setting

$$y_j^* = \frac{\alpha_j}{1 + \sum_{i \neq j_0} \alpha_j}, \qquad y_{j_0}^* = \frac{1}{1 + \sum_{j \neq j_0} \alpha_j}, \qquad Y^* = (y_1^*, \ldots, y_n^*),$$

we obtain that $A_{i \cdot} Y^{*T} \leqslant 0 = v$ for all $i = 1, \ldots, m$. Hence the strategy $Y^*$ for player II is an optimal strategy, and moreover $y_{j_0}^* > 0$. (We thus conclude that the *assumption* of the theorem does not apply in this case.)

Therefore $-A_{\cdot j_0} \notin C$. Then in view of Farkas's lemma proved in Section 1.20.9, there exists a vector $W = (w_1, \ldots, w_n)$ such that

$$WE^{(i)T} \geqslant 0, \qquad i = 1, \ldots, m, \tag{20.13}$$

$$WA_{\cdot j} \geqslant 0, \qquad j \neq j_0, \tag{20.14}$$

but

$$-WA_{\cdot j_0} < 0. \tag{20.15}$$

Inequality (20.13) implies that $w_i \geqslant 0$ and (20.15) yields that $W \neq 0$. Hence the sum of the components of vector $W$ is positive and we can divide, termwise, inequalities (20.14) and (20.15) by this sum.

Setting

$$x_i^* = \frac{w_i}{\sum_{i=1}^{m} w_i}, \qquad X^* = (x_1^*, \ldots, x_m^*),$$

we obtain

$$X^* A_{\cdot j} \geqslant 0, \qquad X^* A_{\cdot j_0} > 0.$$

Hence strategy $X^*$ is an optimal strategy for player I and the required condition (20.11) is satisfied. □

## 1.21 Domination of strategies

*1.21.1* **Definition.** Given a matrix game with the payoff matrix **A**, we say that strategy $X'$ for player I *dominates* his or her strategy $X''$ (and strategy $X''$ is *dominated* by strategy $X'$) if for any pure strategy $j$ of player II

$$X' A_{\cdot j} \geqslant X'' A_{\cdot j}.$$

Analogously, strategy $Y'$ for player II *dominates* his or her strategy $Y''$ if for any pure strategy $i$ for player I

$$A_{i \cdot} Y'^T \leqslant A_{i \cdot} Y''^T.$$

In particular, the pure strategy $i'$ for player I dominates his or her pure strategy $i''$ if for any $j$

$$a_{i'j} \geqslant a_{i''j},$$

and the pure strategy $j'$ for player II dominates his or her pure strategy $j''$ if for any $i$

$$a_{ij'} \leqslant a_{ij''}.$$

Note that the dominance relation refers to the strategies of the *same* player.

*1.21.2* **Definition.** In a matrix game with the payoff matrix **A**, a strategy $X'$ for player I *strictly dominates* his or her strategy $X''$ (and the strategy $X''$ is *strictly dominated* by strategy $X'$) if for any $j$

$$X'A_{\cdot j} > X''A_{\cdot j}.$$

Strategy $Y'$ of player II *strictly dominates* his or her strategy $Y''$ if for any $i$

$$A_{i\cdot}Y' < A_{i\cdot}Y''.$$

The notion of strict dominance is similarly defined for pure strategies.

*1.21.3.* The players in a game need not use their dominated strategies. A more precise meaning of this general assertion is provided by the following theorems.

**Theorem.** *If in a given game a strategy $Z'$ for one of the players dominates his or her strategy $Z''$ and the strategy $Z''$ is optimal, then strategy $Z'$ is also optimal.*

PROOF. For definiteness let the strategies in question be those for player I. Since $Z''$ is an optimal strategy for this player, we have, in view of the theorem in Section 1.16.1, that the value of the game $v$ satisfies $v = \min_j Z''A_{\cdot j}$ so that

$$v \leqslant Z''A_{\cdot j} \quad \text{for any } j. \tag{21.1}$$

Next, in view of the dominance relation,

$$Z''A_{\cdot j} \leqslant Z'A_{\cdot j} \quad \text{for any } j, \tag{21.2}$$

and, therefore,

$$v \leqslant Z'A_{\cdot j} \quad \text{for any } j, \tag{21.3}$$

and the theorem in Section 1.20.5 implies that strategy $Z'$ is an optimal strategy. □

*1.21.4* **Theorem.** *If in a certain game strategy $Z''$ for one of the players is strictly dominated by his or her strategy $Z'$, then $Z''$ is not an optimal strategy.*

Proof. We shall again consider, for definiteness, strategies for player I.

The strict dominance stipulated in the theorem implies that

$$Z''A_{\cdot j} < Z'A_{\cdot j} \quad \text{for any } j,$$

hence (cf. Section 1.6.3)

$$\min_j Z''A_{\cdot j} < \min_j Z'A_{\cdot j}.$$

However, in view of the theorem in Section 1.16.1 $\min_j Z'A_{\cdot j} \leqslant v$ so that

$$\min_j Z''A_{\cdot j} < v \quad \text{for any } j.$$

The same theorem also implies that $Z''$ is not an optimal strategy. □

*1.21.5.* In the case of pure dominated strategies the following result is indicative.

**Theorem.** *If a pure strategy k for a player is dominated by his (pure or mixed) strategy Z, which is different from k, then there exists an optimal strategy $Z^0$ for this player such that the probability is zero that k is contained in $Z^0$.*

Proof. We shall prove this theorem for the case of player I. Let $Z = (z_1, \ldots, z_m)$. Consider the vector $Z' = (z'_1, \ldots, z'_m)$ where

$$z'_i = \begin{cases} \dfrac{z_i}{1 - z_k} & \text{for } i \neq k, \\ 0 & \text{for } i = k. \end{cases} \tag{21.4}$$

Clearly, $z'_i \geqslant 0$ for all $i$ and

$$\sum_{i=1}^{m} z'_i = \sum_{i \neq k} \frac{z_i}{1 - z_k} = \frac{1}{1 - z_k} \sum_{i \neq k} z_i = \frac{1}{1 - z_k}(1 - z_k) = 1,$$

so that vector $Z'$ is a mixed strategy for this player. The dominance relation stipulated in the theorem implies that $ZA_{\cdot j} \geqslant a_{kj}$ for any $j$. This inequality can be rewritten as

$$\sum_{i=1}^{m} z_i a_{ij} \geqslant \sum_{i=1}^{m} z_i a_{kj}.$$

Subtracting the common term $z_k a_{kj}$ from both sides and dividing both sides of the inequality by the (positive!) number $\sum_{i \neq k} z_i = 1 - z_k$, we have

$$\sum_{i \neq k} z'_i a_{ij} = \sum_{i=1}^{m} z'_i a_{ij} \geqslant a_{kj} \quad \text{for any } j, \tag{21.5}$$

i.e. $Z'A_{\cdot j} \geqslant a$ for any $j$.

Let $z^* = (z_1^*, \ldots, z_m^*)$ be an optimal strategy of the player that possibly contains the pure strategy $k$ with a nonzero probability (i.e. $z_k^* > 0$).

Consider the vector $Z^0=(z_1^0,\ldots,z_m^0)$ such that

$$z_i^0=\begin{cases} z_i^*+z_k^*z_i', & \text{if } i\neq k,\\ 0, & \text{if } i=k.\end{cases}$$

Clearly, $z_i^0 \geqslant 0$ and

$$\sum_{i=1}^m z_i^0=\sum_{i\neq k} z_i^*+z_k^*\sum_{i=1}^m z_i'=\sum_{i\neq k} z_i^*+z_k^*=\sum_{i=1}^m z_i^*=1,$$

so that $Z^0$ is indeed a mixed strategy. Utilization of this strategy can be interpreted as utilization of strategy $Z^*$ provided the latter is not "realized" as the pure strategy $k$, and if $k$ is realized, then the mixed strategy $Z'$ will be used. In other words, the pure strategy $k$ appears in the mixed strategy $Z^0$ with zero probability. We need only show that $Z^0$ is an optimal strategy.

For any $j$ we have

$$Z^0A_{\cdot j}=\sum_{i=1}^n z_i^0a_{ij}=\sum_{i\neq k}(z_i^*+z_k^*z_i')a_{ij}=\sum_{i\neq k} z_i^*a_{ij}+z_k^*\sum_{i\neq k} z_i'a_{ij},$$

or taking into account that $z_k'=0$ [cf. (21.4)] we obtain

$$Z^0A_{\cdot j}=\sum_{i\neq k} z_i^*a_{ij}+z_k^*\sum_{i=1}^m z_i'a_{ij}.$$

Now in view of (21.5) and the fact that the numbers $z_k^*$ are nonnegative,

$$Z^0A_{\cdot j}\geqslant \sum_{i\neq k} z_i^*a_{ij}+z_k^*a_{kj}=\sum_{i=1}^m z_i^*a_{ij}=Z^*A_{\cdot j}.$$

Since this inequality is valid for all $j$, $Z^0$ indeed dominates strategy $Z^*$. However, by assumption, strategy $Z^*$ is an optimal strategy. Hence, in view of the theorem in Section 1.21.3, so is $Z^0$ and the assertion is verified. □

*1.21.6* **Theorem.** *If a pure strategy k for a player is strictly dominated by a (pure or mixed) strategy Z, then in any optimal strategy* $Z^*=(z_1^*,\ldots,z_m^*)$ *for this player* $z_k^*=0$.

PROOF. In view of the dominance stipulated in the theorem, we have for each $j$

$$a_{kj}<ZA_{\cdot j}.$$

Choosing some optimal strategy $Y^*$ for player II and passing over in the above inequality to mixed strategies, we obtain

$$A_{k\cdot}Y^{*T}<Z\mathbf{A}Y^{*T}. \tag{21.6}$$

Since $Y^*$ is optimal we have

$$Z\mathbf{A}Y^{*T}\leqslant v;$$

this, together with (21.6), yields

$$A_{k\cdot}Y^{*T} < v.$$

To complete the proof one need only utilize the second part of the theorem in Section 1.20.8. □

## 1.22 Diagonal games

*1.22.1* **Definition.** A matrix game with a payoff matrix of the form

$$\mathbf{A} = \begin{bmatrix} a_1 & 0 & \dots & 0 \\ 0 & a_2 & \dots & 0 \\ \vdots & & & \\ 0 & 0 & \dots & a_n \end{bmatrix},$$

where $a_i > 0$ $(i = 1, \dots, n)$ is called a *diagonal* game.

A diagonal game can be interpreted as the following "search" game. Player II hides an object in one of the $n$ possible locations so that the object of value $a_j$ is hidden in location $j$. Player I searches for the object in one of the possible locations. If he or she finds it, the payoff is the value of the object; if he or she does not, the payoff is zero.

*1.22.2* **Lemma.** *Any diagonal game has a positive value.*

PROOF. Choose the following strategy for player I:

$$X = \left(\frac{1}{n}, \frac{1}{n}, \dots, \frac{1}{n}\right).$$

Then

$$XA_{\cdot j} = \frac{1}{n}a_j > 0, \qquad j = 1, \dots, n;$$

hence, in view of the theorem in item 1 of Section 20, $v > 0$. □

*1.22.3* **Lemma.** *If $X^* = (x_1^*, \dots, x_n^*)$ is an optimal strategy for player* I *in a diagonal game, then all the components of $X^*$ are positive.*

PROOF. Assume that some $x_i^* = 0$. Choosing strategy $j = i$ for player II we obtain

$$v \leqslant X^*A_{\cdot i} = x_1 \cdot 0 + \cdots + x_{i-1} \cdot 0 + 0 \cdot a_i + x_{i+1} \cdot 0 + \cdots + x_n \cdot 0 = 0,$$

i.e. the value of the game is nonpositive, which contradicts the lemma in Section 1.22.2.

*1.22.4.* We now proceed to determine optimal strategies for players in a diagonal game. Since the components of the optimal strategies for player I are all positive, we have in view of the theorem in 1.20.8 that for an arbitrary optimal strategy $Y^*$ of player II

$$A_{i\cdot} Y^{*T} = v \qquad i = 1, \ldots, n.$$

However,

$$A_{i\cdot} Y^{*T} = a_i y_i^*;$$

hence,

$$a_i y_i^* = v.$$

Since $a_j > 0$ the last equation can be written as

$$y_i^* = v / a_i. \tag{22.1}$$

Summing (22.1) with respect to $i$ and taking into account that

$$\sum_{i=1}^{n} y_i^* = 1$$

we have

$$v = \frac{1}{\sum_{j=1}^{n} 1/a_j}.$$

Substituting this expression for $v$ into (22.1) we finally obtain

$$y_i^* = \frac{1}{a_i} \frac{1}{\sum_{j=1}^{n} 1/a_j}. \tag{22.2}$$

Formula (22.2) shows that all components $y_i^*$ of an optimal strategy for player II are positive. Hence to determine an optimal strategy for player I one can again apply the theorem in Section 1.20.8. According to this theorem, for any optimal strategy $X^* = (x_1^*, \ldots, x_n^*)$ for player I and any $j = 1, \ldots, n$ we have

$$X^* A_{\cdot j} = v = \frac{1}{\sum_{j=1}^{n} 1/a_j}. \tag{22.3}$$

From (22.3) we easily see that

$$x_j^* = \frac{1}{a_j} \cdot \frac{1}{\sum_{j=1}^{n} 1/a_j}. \tag{22.4}$$

Hence, player I in a diagonal game also has a unique optimal strategy.

*1.22.5.* The result given by formula (22.4) seems to be paradoxical at first glance: it shows that player I ought to search with higher probability for less valuable objects. However, if we take into account the optimal strategy of player II given by (22.2), this paradoxical conclusion is resolved.

Indeed, formula (22.2) indicates that using optimal behavior, player II more often than not hides the less valuable objects. It is therefore only appropriate for player I to search for these objects.

Observe also that in a diagonal game, the vectors of optimal strategies for players I and II are the same. However, it would seem unreasonable to suggest that the optimal strategies of these players coincide. The optimal strategies express completely different actions for each one of the players. What is equal is only the probabilities with which the players revert to the same place.

## 1.23 Sets of optimal strategies in a matrix game

*1.23.1.* The set of all optimal strategies for player I in a matrix game with the payoff matrix **A** is commonly denoted by $T_1(\mathbf{A})$, and the set of all optimal strategies for player II in this game by $T_2(\mathbf{A})$.

*1.23.2* **Theorem.** *In any matrix game with a payoff matrix A, each one of the sets* $T_1(\mathbf{A})$ *and* $T_2(\mathbf{A})$ *is nonempty, convex, closed, and bounded.*

We shall prove this theorem for the set $T_1(\mathbf{A})$ only.

The fact that the set $T_1(\mathbf{A})$ is *nonempty* follows from the existence of optimal strategies, which was established in Section 1.14.

Further the set $T_1(\mathbf{A})$ in view of Section 1.20.5 and of the definition of optimal strategy is the intersection of a finite number of closed half-spaces that consist of vectors $X$ defined by inequalities

$$v \leqslant XA_{\cdot j} \qquad j=1,\ldots,n.$$

This intersection is *convex* and *closed.*

Finally, since the set $T_1(\mathbf{A})$ is contained in the fundamental simplex of strategies $S_m$, which is a bounded set, $T_1(\mathbf{A})$ being a subset of a bounded set is also *bounded.*

## 1.24 An example: 3×3 games

*1.24.1.* 3×3 games occur in many applications. Therefore it is useful to know how to solve this type of game directly without resorting to the general methodology (such as the simplex method of linear programming, cf. Section 1.26) or to general computational techniques.

It is hardly possible to analyze these games in such complete detail as was done in Section 1.14 for the 2×2 games. The number of possible particular cases is substantially larger in this case. Therefore, we shall confine ourselves to the general description of a method for solving a 3×3 game without providing actual computational formulas in various particular cases.

Let a game with a 3×3 payoff matrix **A** be given. The idea behind its solution is based on the following considerations.

*1.24.2.* First, as it was already mentioned on several occasions, the value $v$ of any matrix game with a payoff matrix **A** is equal to

$$v = \max_{X} \min_{j} XA_{\cdot j}, \tag{24.1}$$

where the outer maximum is attained at the optimal strategies for player I and only at these strategies.

In our case $X$ is a vector of the form $(x_1, x_2, x_3)$, where

$$x_1, x_2, x_3 \geqslant 0, \tag{24.2}$$

$$x_1 + x_2 + x_3 = 1. \tag{24.3}$$

Denote the set of all these vectors by $S$ (cf. Section 1.10.1).

Equation (24.1) can now be rewritten explicitly in the form

$$v = \max_{X \in S} \min \begin{Bmatrix} x_1 a_{11} + x_2 a_{21} + x_3 a_{31} \\ x_1 a_{12} + x_2 a_{22} + x_3 a_{32} \\ x_1 a_{13} + x_2 a_{23} + x_3 a_{33} \end{Bmatrix} = \max_{X \in S} \min \begin{Bmatrix} XA_{\cdot 1} \\ XA_{\cdot 2} \\ XA_{\cdot 3} \end{Bmatrix}. \tag{24.4}$$

Consider the equalities

$$XA_{\cdot 1} = XA_{\cdot 2}, \tag{24.5}$$

$$XA_{\cdot 2} = XA_{\cdot 3}, \tag{24.6}$$

$$XA_{\cdot 3} = XA_{\cdot 1}. \tag{24.7}$$

Each of these equalities [provided $X = (x_1, x_2, x_3)$ also satisfies (24.3)] is either an identity or an equation of a line that subdivides the plane (24.3) into two half-planes; moreover, in each of these half-planes one of the linear forms appearing at one side of equalities (24.5)–(24.7) takes on larger values than the second appearing at the other side. Carrying out all these subdivisions of the plane (24.3) and comparing the values of the forms on the corresponding half-planes, we can determine the regions of the plane where a given form (i.e. the first, second, or third) takes on the smallest value (or two of the given forms take on the smallest value). Denote the region in which the form $XA_{\cdot j}$ has the minimal value by $K_j$.

This implies that

$$\min_{j=1,2,3} XA_{\cdot j} = \begin{cases} XA_{\cdot 1} & \text{for } X \in K_1, \\ XA_{\cdot 2} & \text{for } X \in K_2, \\ XA_{\cdot 3} & \text{for } X \in K_3. \end{cases}$$

In general, some of the sets $K_1$, $K_2$, and $K_3$ may coincide and some may also be empty.

If the system of equations (24.3)–(24.6) [(24.7) is a corollary to (24.5) and (24.6)] possesses a unique solution, then regions $K_1$, $K_2$, and $K_3$ are angular regions as indicated in Figure 12. If, however, this system possesses infinitely many solutions, these regions become strip regions or half-planes (see Figure 13).

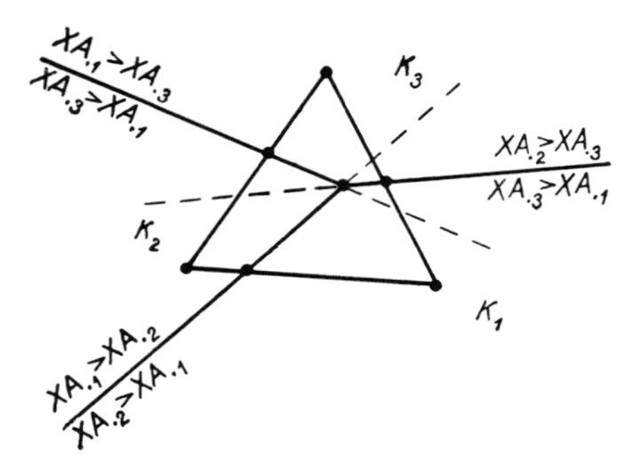


Figure 12

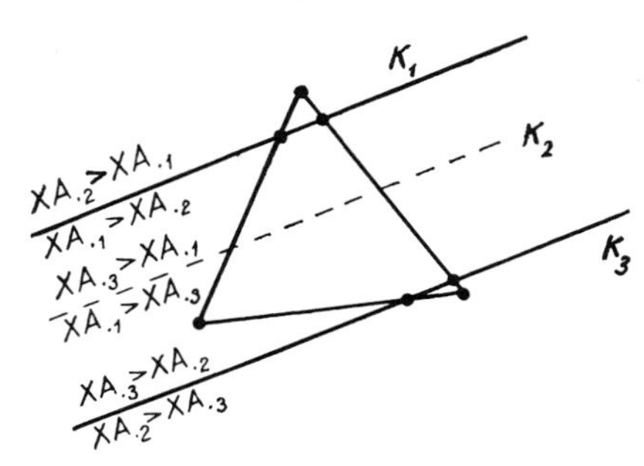


Figure 13

Using the notation introduced above, we can rewrite (24.1) or equivalently (24.4) in the form

$$
\begin{aligned}
v &= \max_{X \in S} \min_{j=1,2,3} XA_{\cdot j} \\
&= \max\left\{ \max_{X \in S \cap K_1} XA_{\cdot 1}, \max_{X \in S \cap K_2} XA_{\cdot 2}, \max_{X \in S \cap K_3} XA_{\cdot 3} \right\}. \qquad (24.8)
\end{aligned}
$$

*1.24.3.* We need now to determine the interior maxima in (24.8) and compare them with one another. Observe that each of the sets $S \cap K_j$ is an intersection of a triangle with an angular region, a strip region or a half-plane. Therefore, a set $S \cap K_j$ is either empty, or contains a single point, or is a closed interval, or a triangle or, finally, a convex quadrangle.

However, a maximum of a linear form on a segment, triangle, or convex quadrangle can be attained only at a vertex (the proof of this assertion is actually the same as the proof of the lemma in Section 1.10.2). Therefore, to determine a maximum in (24.8) it is sufficient to find the vertices of the polygons $S \cap K_j$, to evaluate the values of the corresponding form at these vertices, and to compare them.

The vertices of each one of the polygons $S \cap K_j$ are either the vertices of $S$ belonging to $K_j$, or a vertex of $K_j$ (provided it exists) belonging to $S$, or the points of intersection of the sides of $K_j$ with the sides of $S$. All these points can be enumerated easily.

The vertices of the triangle $S$ are the points

$$1=(1,0,0), \qquad 2=(0,1,0), \qquad 3=(0,0,1).$$

Whether a particular point $X$ belongs to a *given* set $K_j$ can be directly checked by determining whether $XA_{\cdot j}$ is the smallest among the forms $XA_{\cdot 1}$, $XA_{\cdot 2}$, and $XA_{\cdot 3}$ (or at least not larger than the two others).

Furthermore, if one of the sets $K_j$ has a vertex, we are then dealing with the case represented in Figure 12. Moreover, all the sets $K_j$ ought to have a common and unique vertex. Whether this vertex belongs to the set $S$ is checked by verifying that conditions (24.2) and (24.3) are satisfied for the coordinates of this vertex.

Finally, the equation of a side of $K_j$ is given by

$$XA_{\cdot j_1}=XA_{\cdot j_2}, \tag{24.9}$$

and that of $S$ by

$$x_i=0. \tag{24.10}$$

Combining all the equations of the form (24.9) with the equations (24.10) and taking conditions (24.2) and (24.3) into account, we can determine all the points of this type also.

As it is evident from Figures 12 and 13, the maximal number of points at which the linear forms ought to be evaluated and compared is seven. The total number of combinatorially distinct variants is, however, quite large in this case and we shall not enumerate them.

*1.24.4.* After the value of the game has been determined, there is no difficulty in determining the optimal strategies for player II. We need only find the values of the vector $Y=(y_1,y_2,y_3)$ satisfying the "usual" relations

$$y_1,y_2,y_3 \geqslant 0, \tag{24.11}$$

$$y_1+y_2+y_3=1 \tag{24.12}$$

(the set of all these vectors forms a triangle—a simplex to be denoted by $T$) and such that

$$\max_{i=1,2,3} A_{i\cdot}Y^T=v \tag{24.13}$$

is satisfied.

For this purpose it is sufficient to find those regions $L_1$, $L_2$, and $L_3$ in the plane (24.12) where the forms $A_{1\cdot}Y^T$, $A_{2\cdot}Y^T$, and $A_{3\cdot}Y^T$, respectively, take on the minimal value and determine the intersection of each of these regions with the triangle.

These points of intersection constitute the set of optimal strategies for player II.

*1.24.5.* Clearly, there are many opportunities to simplify and reduce the required arguments and computations in the process of actually determining the value of a game and the corresponding optimal strategies if we use the method described above. (It is, however, hardly desirable to enumerate all these special cases.)

*1.24.6* EXAMPLE. Let

$$\mathbf{A}=\begin{bmatrix}1 & 1 & 2\\ 0 & 2 & 0\\ 2 & 0 & 0\end{bmatrix}.$$

Equation (24.5) becomes in this case

$$XA_{\cdot 1}=x_1+2x_3=x_1+2x_2=XA_{\cdot 2},$$

or $x_3=x_2$.

Analogously, (24.6) can be written as

$$XA_{\cdot 2}=x_1+2x_2=2x_1=XA_{\cdot 3},$$

i.e. $2x_2=x_1$.
Finally, (24.7) becomes

$$XA_{\cdot 3}=2x_1=x_1+2x_3=XA_{\cdot 1},$$

hence $x_1=2x_3$.

Hence the regions $K_1, K_2, K_3$ in this case are as presented in Figure 14. [The coordinates of the point 0 are $(\frac{1}{2},\frac{1}{4},\frac{1}{4})$.] Computing the values of the form $XA_{\cdot 1}$ at the points *1, 2, 0*, we obtain

$$(1,0,0)(1,0,2)^T=1,$$

$$(0,1,0)(1,0,2)^T=0,$$

$$\left(\tfrac{1}{2},\tfrac{1}{4},\tfrac{1}{4}\right)(1,0,2)^T=1,$$

and an evaluation of the value of the form $XA_{\cdot 2}$ at point *3* yields

$$(0,0,1)(1,2,0)^T=0.$$

Consequently, the vectors $(1,0,0)$ and $(\frac{1}{2},\frac{1}{4},\frac{1}{4})$ are actually the optimal strategies for player I. Hence, in view of the discussion in Section 1.24.3,

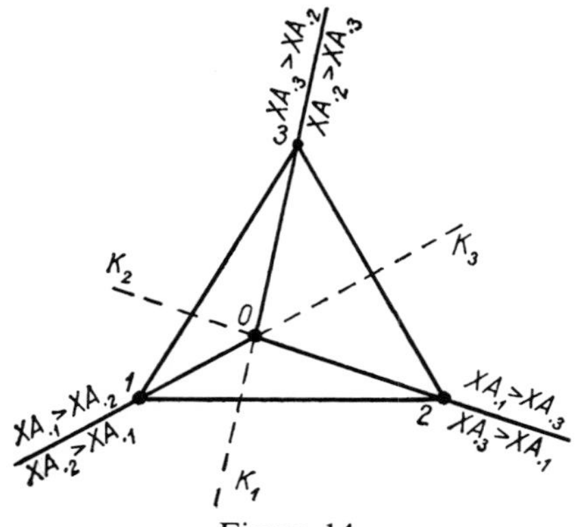


Figure 14

the set of all optimal strategies for player I is the interval joining the corresponding points *1* and *0*. The value of the game is 1.

We now proceed to determine the optimal strategies for player II. Equating the corresponding linear forms we first obtain

$$A_{1.}Y^T = y_1 + y_2 + 2y_3 = 2y_2 = A_{2.}Y^T,$$

whence $1 + y_3 = 2y_2$; next

$$A_{2.}Y^T = 2y_2 = 2y_1 = A_{3.}Y^T,$$

so that $y_2 = y_1$; finally

$$A_{3.}Y^T = 2y_1 = y_1 + y_2 + 2y_3 = A_{1.}Y^T;$$

thus $2y_1 = 1 + y_3$.

Hence the regions $L_1$, $L_2$, and $L_3$ are of the form as presented in Figure 15. We have

$$U = \left(\tfrac{2}{3}, 0, \tfrac{1}{3}\right),$$
$$V = \left(0, \tfrac{2}{3}, \tfrac{1}{3}\right),$$
$$W = \left(\tfrac{1}{2}, \tfrac{1}{2}, 0\right).$$

Computing the values of the form $A_{1.}Y^T$ at the points $U$, $V$, and $W$, we obtain

$$(1,1,2)\left(\tfrac{2}{3}, 0, \tfrac{1}{3}\right)^T = \tfrac{4}{3}, \qquad (1,1,2)\left(0, \tfrac{2}{3}, \tfrac{1}{3}\right)^T = \tfrac{4}{3},$$
$$(1,1,2)\left(\tfrac{1}{2}, \tfrac{1}{2}, 0\right)^T = 1, \qquad (1,1,2)(0,0,1)^T = 2,$$

while computing at points *2* and *1* the values of the forms $A_{2.}Y^T$ and $A_{3.}Y^T$, respectively, we have

$$(0,2,0)(0,1,0)^T = 2, \qquad (2,0,0)(1,0,0)^T = 2.$$

Here the only optimal strategy for player II is the point $W = (\frac{1}{2}, \frac{1}{2}, 0)$.

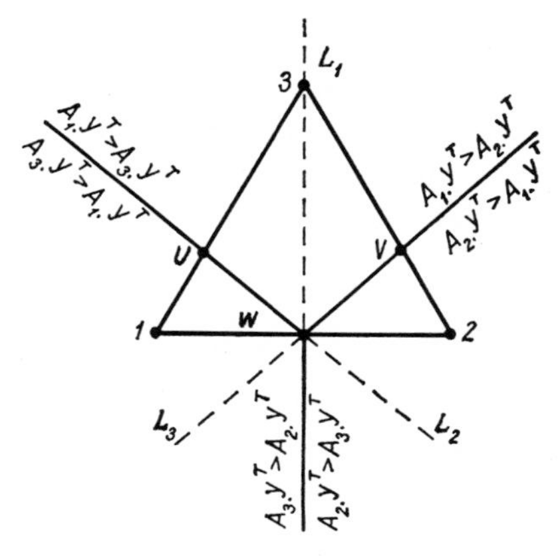


Figure 15

## 1.25 Symmetric games

*1.25.1* **Definition.** A matrix game with a skew-symmetric payoff matrix, i.e. with a payoff matrix $\mathbf{A}$ satisfying $\mathbf{A} = -\mathbf{A}^T$, is called a *symmetric* game.

*1.25.2* **Theorem.** *In a symmetric game with a payoff matrix* $\mathbf{A}$,

$$T_1(\mathbf{A}) = T_2(\mathbf{A}), \qquad v(\mathbf{A}) = 0.$$

PROOF. Let $X^* \in T_1(\mathbf{A})$ and $Y^* \in T_2(\mathbf{A})$. This implies that

$$X\mathbf{A}Y^{*T} \leqslant X^*\mathbf{A}Y^{*T} \leqslant X^*\mathbf{A}Y^T \quad \text{for all } Y, X \in S_m.$$

Transposing all the matrices and vectors appearing in this relation we obtain

$$Y^*\mathbf{A}^T X^T \leqslant Y^*\mathbf{A}^T X^{*T} \leqslant Y\mathbf{A}^T X^{*T} \quad \text{for all } Y, X \in S_m.$$

Multiplying all the sides of the last inequality by $-1$, we have

$$Y^*(-\mathbf{A}^T)X^T \geqslant Y^*(-\mathbf{A}^T)X^{*T} \geqslant Y(-\mathbf{A}^T)X^{*T} \quad \text{for all } Y, X \in S_m,$$

and taking into account that the game is symmetric,

$$Y^*\mathbf{A}X^T \geqslant Y^*\mathbf{A}X^{*T} \geqslant Y\mathbf{A}X^{*T} \quad \text{for all } Y, X \in S_m,$$

i.e. $Y^* \in T_1(\mathbf{A})$ and $X^* \in T_2(\mathbf{A})$, which shows that $T_1(\mathbf{A}) = T_2(\mathbf{A})$.

Moreover,

$$v(\mathbf{A}) = X^*\mathbf{A}Y^{*T} = X^*(-\mathbf{A}^T)Y^{*T} = -X^*\mathbf{A}^T Y^{*T} = -Y^*\mathbf{A}X^{*T} = -v(\mathbf{A}),$$

whence $v(\mathbf{A}) = 0$. □

## 1.26 Matrix games and linear programming

*1.26.1.* A solution of a matrix game can be reduced to a solution of a standard linear programming problem.

Consider a game with a $m \times n$ payoff matrix $\mathbf{A}$. Without loss of generality, we may assume that all the entries in this matrix are positive (otherwise, we may add a sufficiently large number to all the entries, and the new game will be strategically equivalent to the initial game).

Let $X$ be an arbitrary strategy for player I in this game. Set $v_X = \min_j XA_{\cdot j}$. Since all the entries of $\mathbf{A}$ are positive, so is $v_X$ for any strategy $X$. Thus,

$$v_X \leqslant XA_{\cdot j}, \quad j = 1, \ldots, n. \tag{26.1}$$

It follows from the theorem in Section 1.20.1 that $v_X \leqslant v(\mathbf{A})$ [where $v(\mathbf{A})$ is the value of the game] and hence, in view of the theorem in Section 1.20.5, the equality $v_X = v(\mathbf{A})$ is a necessary and sufficient condition for $X$ to be an optimal strategy. Hence the optimality of strategy $X$ is equivalent to the equality

$$v_X = \max_X v_X = v(\mathbf{A}). \tag{26.2}$$

Consider now the vector

$$\tilde{X}=(1/v_X)X.$$

Inequalities (26.1) yield

$$\tilde{X}A_{\cdot j}\geqslant 1 \tag{26.3}$$

The fact that $X$ is a strategy means that

$$\tilde{X}J_m^T=1/v_X, \tag{26.4}$$

where $J_m$ is the vector $(1,\dots,1)$ defined in Section 1.9.3 and, moreover,

$$\tilde{X}\geqslant \mathbf{0}. \tag{26.5}$$

Equations (26.2) and (26.4) imply that strategy $X$ is optimal if and only if

$$\tilde{X}J_m^T\to\min. \tag{26.6}$$

(i.e. it is required to minimize the linear function $\tilde{X}J_m^T$).

The conditions (26.3), (26.5), and (26.6) define a standard linear program problem. The value of the minimum of $\tilde{X}J_m^T$ (i.e. the "value" of the problem) is the reciprocal of the value of the game and every solution of this problem, upon appropriate normalization (i.e. after division by the value of the problem) is an optimal strategy for player I.

*1.26.2.* Analogously, we define for each strategy $Y$ of player II the quantity

$$v_Y=\max_i A_{i\cdot}Y^T,$$

and setting

$$\tilde{Y}=(1/v_Y)Y,$$

we arrive at the linear programming problem

$$\begin{aligned}A_{i\cdot}\tilde{Y}^T\leqslant 1, \quad i=1,\dots,m,\\ \tilde{Y}\geqslant \mathbf{0},\\ J_n\tilde{Y}^T\to\max,\end{aligned}$$

which is dual to the linear programming problem defined by (26.3), (26.5), and (26.6). Consequently, the solution of this dual linear programming problem and only these solutions are the optimal strategies for player II in the game with matrix **A**.

*1.26.3.* In preceding items we have shown that a solution of the matrix game can be reduced to a pair of dual linear programming problems. Now we shall prove that conversely, if a pair of dual linear programming problems possesses solutions, then the set of these solutions is completely described by the set of solutions of a certain matrix game. This will indicate that the theory of matrix games is in a certain sense equivalent to the theory of standard linear programming problems.

Let a pair of dual problems of linear programming be given by

$$X\mathbf{A} \leqslant B, \tag{26.7}$$

$$X \geqslant \mathbf{0}, \tag{26.8}$$

$$XC^T \to \max$$

and

$$\mathbf{A}Y^T \geqslant C^T, \tag{26.9}$$

$$Y \geqslant \mathbf{0}, \tag{26.10}$$

$$BY^T \to \min,$$

where $X$ and $C$ are $m$-dimensional vectors, $Y$ and $B$ are $n$-dimensional vectors, and $\mathbf{A}$ is an $m \times n$ matrix.

*1.26.4.* Multiplying inequality (26.7) termwise on the right by $Y^T$, we obtain

$$X\mathbf{A}Y^T \leqslant BY^T \tag{26.11}$$

and multiplying inequality (26.8) termwise on the left by $X$ we have

$$X\mathbf{A}Y^T \geqslant XC^T. \tag{26.12}$$

It follows from (26.11) and (26.12) that

$$XC^T \leqslant BY^T \tag{26.13}$$

for any $X$ and $Y$ satisfying inequalities (26.7)–(26.10). Consequently, if an equality is attained in (26.13), i.e. if

$$XC^T = BY^T,$$

then $XC^T$ takes on the largest possible value subject to the restrictions (26.7) and (26.8), and $BY^T$ takes on the smallest possible value subject to the restrictions (26.9) and (26.10). In other words $X$ and $Y$ are indeed the optimal solutions for the linear programming problems under consideration.

The argument presented in this item is often referred to as the "duality theorem" of linear programming.

*1.26.5.* Given the pair of linear programming problems (26.7), (26.8) and (26.9), (26.10), we shall associate with this pair a $(m+n+1)\times(m+n+1)$ matrix $\mathbf{M}$ and write it in the following "block" form:

$$\mathbf{M} = \begin{Vmatrix} \mathbf{0} & -\mathbf{A} & C^T \\ \mathbf{A}^T & \mathbf{0} & -B^T \\ -C & B & 0 \end{Vmatrix}.$$

This matrix is a skew-symmetric one; hence, a matrix game with this matrix is symmetric (Section 1.25). Thus the value of the game is zero and the players possess the same optimal strategies. In what follows we shall call these strategies optimal without specifically indicating the players or the game itself.

Let $Z=(U,V,t)$ be an optimal strategy; here $U$ is an $m$-vector, $V$ is an $n$-vector, and $t$ is a scalar. The fact that $Z$ is a strategy implies that

$$Z \geqslant \mathbf{0}, \qquad ZJ^T_{m+n+1}=1, \tag{26.14}$$

and the fact that it is an optimal strategy yields

$$Z\mathbf{M} > \mathbf{0}. \tag{26.15}$$

*1.26.6.* A detailed description of a relation between the solvability of a pair of dual linear programming problems and the determination of their strategies on one hand and solutions of matrix games on the other, is presented by the following theorem.

**Theorem.** *Let the following pair of dual linear programming problems,*

$$\left.\begin{aligned} &X\mathbf{A} \leqslant B,\\ &X \geqslant \mathbf{0},\\ &XC^T \to \max, \end{aligned}\right\} \quad \text{"direct problem"}$$

$$\left.\begin{aligned} &\mathbf{A}Y^T \geqslant C^T,\\ &Y \geqslant \mathbf{0},\\ &BY^T \to \min \end{aligned}\right\} \quad \text{"dual problem"}$$

*and the associated matrix* $\mathbf{M}$ *defined above be given. Then*:

(1) *if* $Z=(U,V,t)$ *is an optimal strategy with* $t>0$, *then*

$$X=\frac{1}{t}U, \qquad Y=\frac{1}{t}V \tag{26.16}$$

*are the optimal solutions of the direct and dual linear programming problems, respectively*;

(2) *if for any optimal strategy* $Z=(U,V,t)$ *we have* $t=0$, *then both linear programming problems possess no optimal solution*;

(3) *If* $X$ *and* $Y$ *are solutions of the direct and dual linear programming problems, respectively, and*

$$t=\frac{1}{XJ^T_m+YJ^T_n+1}, \tag{26.17}$$

*then*

$$Z=(tX,tY,t) \tag{26.18}$$

*is an optimal strategy*.

Proof. (1) Write the relations (26.14) and (26.15) using the "block" form:

$$(U,V,t) \geqslant \mathbf{0},$$

$$(U,V,t)J^T_{m+n+1}=1,$$

$$(U,V,t)\begin{bmatrix} \mathbf{0} & -\mathbf{A} & C^T \\ \mathbf{A}^T & 0 & -B^T \\ -C & B & 0 \end{bmatrix} \geqslant 0.$$

Carrying out the corresponding block multiplication we get

$$U \geqslant \mathbf{0}, \qquad V \geqslant \mathbf{0}, \qquad t \geqslant 0, \tag{26.19}$$

$$UJ_m^T + VJ_n^T + t = 1, \tag{26.20}$$

$$\mathbf{A}V^T - tC^T \geqslant \mathbf{0}, \tag{26.21}$$

$$-U\mathbf{A} + tB \geqslant \mathbf{0}, \tag{26.22}$$

$$UC^T - BV^T \geqslant 0. \tag{26.23}$$

It is assumed that $t > 0$. Consequently, all the numbers and the relations can be divided by $t$ (those involving matrices or vectors are divided termwise). Dividing by $t$ the vectors $U$ and $V$, we obtain (26.16); dividing (26.22), (26.21), and (26.19), we obtain the statements of the direct linear programming problem and the dual linear programming problem as stipulated in the assumptions of the theorem.

Hence in view of the "duality theorem" of linear programming (cf. Section 1.26.4), (26.13) should be valid. However, (26.23) implies that $XC^T \geqslant BY^T$ so that (26.13) holds with an equality sign. It thus follows (recall the argument in Section 1.26.4) that $X$ and $Y$ form a pair of optimal solutions for the pair of dual linear programming problems.

(2) Now let equality $t = 0$ be valid for any optimal strategy $Z = (U, V, t)$. In view of the theorem in Section 1.20.10 this implies a strict inequality in (26.23), i.e. $UC^T - BV^T > 0$ for some optimal strategy $Z = (U, V, t)$.

Thus the system (26.21)–(26.23) may be rewritten as

$$\mathbf{A}V^T \geqslant \mathbf{0}, \tag{26.24}$$

$$U\mathbf{A} \leqslant \mathbf{0}, \tag{26.25}$$

$$UC^T > BV^T. \tag{26.26}$$

There are two possibilities as to the sign of the quantities in (26.26):

(a) First, let us assume that

$$UC^T > 0. \tag{26.27}$$

Choose $\alpha > 0$, a solution $X$ of the linear programming problem, and consider the vector $X + \alpha U$. In view of (26.25) and the inequality that describes the restrictions in the direct linear programming problem, we have

$$(X + \alpha U)\mathbf{A} = X\mathbf{A} + \alpha U\mathbf{A} \leqslant B,$$

and in view of (26.19) and the fact that vector $X$ is nonnegative,

$$X + \alpha U \geqslant \mathbf{0}.$$

Hence $X + \alpha U$ is one of the "feasible" solutions of the above-stated direct linear programming problem (i.e. $X + \alpha U$ is a vector that satisfies all the restrictions). However, as $\alpha$ increases indefinitely so is the product

$$(X + \alpha U)C^T = XC^T + \alpha UC.$$

This shows that there is no optimal solution of the direct linear programming problem.

Since the left-hand side of (26.13) may become larger than any arbitrarily chosen fixed number, there is no $Y$ such that this inequality is satisfied for all $X$. Consequently, there is no *feasible* solution for the dual problem either, so that a fortiori there is no optimal solution for this problem.

(b) Assume now that

$$BV^T < 0.$$

Analogous considerations as in case (a) show that there is no optimal solution for the dual problem and also no feasible solution for the direct problem.

(3) Now let $X$ and $Y$ be optimal solutions of the given pair of dual problems. Choose $t$ and $Z$ in accordance with (26.17) and (26.18). Clearly, $Z \geqslant \mathbf{0}$ and $ZJ_{m+n+1}^T = 1$, so that $Z$ is indeed a strategy in a game with the payoff matrix $\mathbf{M}$. Moreover,

$$\begin{aligned} Z\mathbf{M} &= (tX, tY, t)\begin{bmatrix} \mathbf{0} & -\mathbf{A} & C^T \\ \mathbf{A}^T & \mathbf{0} & -B^T \\ -C & B & 0 \end{bmatrix} \\ &= t(Y\mathbf{A}^T - C, -X\mathbf{A} + B, XC^T - YB^T). \end{aligned}$$

In view of (26.9), (26.7), and (26.13) all the components of vector $Z$ are nonnegative, i.e. are not less than the value of the game. Therefore $Z$ is an optimal strategy of the game under consideration and the required assertion is proved. □

The theorem just proved implies that any method for solving a linear programming problem can easily be adapted to solve a matrix game.

# 2 Infinite antagonistic games

## 2.1 Introduction and motivation

*2.1.1.* In many of the natural sciences (and recently in various social sciences) it is quite common to replace the study of finite sets with a large number of elements by (corresponding) infinite sets. This method allows us to apply the powerful techniques of mathematical analysis to a great variety of problems. Therefore, if we take the case of a game with a large number of strategies for a player then—regardless of whether he or she actually possesses a finite or infinite number of strategies—it is methodologically natural as well as practically useful to assume that the set of strategies for this player is infinite.

An infinite antagonistic game is an antagonistic game in which at least one of the players possesses an infinite set of strategies.

Throughout this chapter the following notation is utilized: $\mathfrak{X}$ and $\mathfrak{Y}$ denote the sets of all the *pure* strategies for players I and II, respectively; $x$ and $y$ (possibly with subscripts) are arbitrary pure strategies for these players, and $X$ and $Y$ (possibly with subscripts) are arbitrary *mixed* strategies for players I and II, respectively.

*2.1.2.* As is the case in any antagonistic game, the optimal behavior of a player is the one that follows the maximin "route" (or principle) of action. It was established in Section 1.7 that this principle is actually realized in a game $\Gamma = \langle \mathfrak{X}, \mathfrak{Y}, H \rangle$ if and only if there exist the mixed extrema

$$\max_{x \in \mathfrak{X}} \inf_{y \in \mathfrak{Y}} H(x,y) \quad \text{and} \quad \min_{y \in \mathfrak{Y}} \sup_{x \in \mathfrak{X}} H(x,y), \tag{1.1}$$

which are moreover equal to each other.

The existence of these extrema for finite antagonistic (matrix) games was demonstrated in Section 1.10. For infinite antagonistic games the situation is generally not as simple.

Consider, for example, the game with $\mathfrak{X}=\mathfrak{Y}=(0,1)$ and the payoff function $H(x,y)=x+y$ (cf. Figure 2, Section 1.5).

If strategies $x=1$ and $y=0$ belonged to the sets of strategies for the players in this game, the situation (1,0) would be the saddle point of the game. [It is at these strategies that the outer extrema are attained in the expression (1.1).]

In that case the value of the game would be the number 1. In actuality, however, the outer extrema in (1.1) are *not* attained and these possibilities are not realized (since $x=1$ and $y=0$ *do not* belong to $\mathfrak{X}$ and $\mathfrak{Y}$, respectively). It is, however, clear that player I, by choosing for his or her strategy a number sufficiently close to unity, will definitely receive a payoff as close to the value of the game as possible. In the same manner, player II, by choosing any number sufficiently close to 0 will assure that his or her loss be arbitrarily close to the value of the game. It thus makes sense, in connection with the game under consideration, to speak about the optimality of strategies "up to an arbitrary $\varepsilon>0$." A precise definition of these notions is given in the next section.

## 2.2 Situations of ε-equilibrium; ε-saddle points and ε-optimal strategies

*2.2.1* **Definition.** Given an antagonistic game Γ, the situation $(x_\varepsilon,y_\varepsilon)$ is called a *situation of ε-equilibrium* if the inequality

$$H(x,y_\varepsilon)-\varepsilon\leqslant H(x_\varepsilon,y_\varepsilon)\leqslant H(x_\varepsilon,y)+\varepsilon \tag{2.1}$$

is valid for any strategies $x$ and $y$ for players I and II, respectively. A point $(x_\varepsilon,y_\varepsilon)$ at which (2.1) is satisfied is also called an *ε-saddle point* of function $H$.

*2.2.2* **Definition.** Strategies $x_\varepsilon$ and $y_\varepsilon$ that form a situation of ε-equilibrium in an antagonistic game are called *ε-optimal strategies*.

This terminology reflects the fact that these strategies are optimal "up to an ε." Namely, while no advantage is gained for a player by deviating from his or her *optimal* strategy, in the same vein, if a player deviates from his or her *ε-optimal* strategy, the payoff can be increased but at most by ε.

## 2.3 ε-optimal strategies and minimaxes

*2.3.1.* A criterion for the existence of an ε-saddle point of a function for any given $\varepsilon>0$ is similar to the criterion for the existence of a saddle point that was derived in Section 1.7.

**Theorem.** (1) *If* $(x_\varepsilon, y_\varepsilon)$ *is an* $\varepsilon$*-saddle point of a function of two variables* $f$, *then*

$$\inf_y f(x_\varepsilon, y) + 2\varepsilon \geqslant \sup_x \inf_y f(x,y), \tag{3.1}$$

$$\sup_x (x, y_\varepsilon) - 2\varepsilon \leqslant \inf_x \sup_y f(x,y). \tag{3.2}$$

(2) *If for any* $\varepsilon > 0$ *the function* $f$ *possesses an* $\varepsilon$*-saddle point, then*

$$\sup_x \inf_y f(x,y) = \inf_y \sup_x f(x,y). \tag{3.3}$$

(3) *If conditions* (3.1), (3.2), *and* (3.3) *are fulfilled, then* $(x_\varepsilon, y_\varepsilon)$ *is a* $4\varepsilon$*-saddle point.*

(4) *If equality* (3.3) *is satisfied, then the function* $f$ *possesses an* $\varepsilon$*-saddle point.*

PROOF. Let $(x_\varepsilon, y_\varepsilon)$ be an $\varepsilon$-saddle point of function $f$. This implies that

$$f(x, y_\varepsilon) - \varepsilon \leqslant f(x_\varepsilon, y_\varepsilon) \leqslant f(x_\varepsilon, y) + \varepsilon. \tag{3.4}$$

Moreover, for any $x$ we obviously have

$$\inf_y f(x,y) - \varepsilon \leqslant f(x, y_\varepsilon) - \varepsilon. \tag{3.5}$$

Equations (3.4) and (3.5) imply that

$$\inf_y f(x,y) - \varepsilon \leqslant f(x, y_\varepsilon) - \varepsilon \leqslant f(x_\varepsilon, y) + \varepsilon.$$

Approaching the supremum with respect to $x$ and the infimum with respect to $y$ in the last inequality, we obtain

$$\sup_x \inf_y f(x,y) - \varepsilon \leqslant \sup_x f(x, y_\varepsilon) - \varepsilon \leqslant \inf_y f(x_\varepsilon, y) + \varepsilon. \tag{3.6}$$

Comparing the extremal parts of inequality (3.6), we immediately obtain (3.1). An analogous argument yields (3.2) and the first assertion of the theorem is thus proved.

Taking the inf with respect to $y$ in the sup $f(x, y_\varepsilon)$ and the sup with respect to $x$ in the inf $f(x_\varepsilon, y)$ we obtain

$$\inf_y \sup_x f(x,y) - \varepsilon \leqslant \sup_x \inf_y f(x,y) + \varepsilon$$

in the right-hand side of inequality (3.6). Since $\varepsilon > 0$ is arbitrary we deduce from the last inequality that

$$\inf_y \sup_x f(x,y) \leqslant \sup_x \inf_y f(x,y).$$

The reverse inequality is, however, always valid (see Section 1.6.4). Hence (3.3) is verified and the proof of the second assertion of the theorem is completed.

Furthermore, it follows from (3.1), (3.2), and (3.3) that

$$\sup_x f(x, y_\varepsilon) - 2\varepsilon \leqslant \inf_y f(x_\varepsilon, y) + 2\varepsilon,$$

i.e. we have for all $x$ and $y$

$$f(x,y_\varepsilon)-2\varepsilon \leqslant f(x_\varepsilon,y)+2\varepsilon. \tag{3.7}$$

In particular, setting $x=x_\varepsilon$, we obtain

$$f(x_\varepsilon,y_\varepsilon) \leqslant f(x_\varepsilon,y)+4\varepsilon \quad \text{for all } y. \tag{3.8}$$

Analogously, setting $y=y_\varepsilon$, we have

$$f(x,y_\varepsilon)-4\varepsilon \leqslant f(x_\varepsilon,y_\varepsilon) \quad \text{for all } x. \tag{3.9}$$

Inequalities (3.8) and (3.9) yield

$$f(x,y_\varepsilon)-4\varepsilon \leqslant f(x_\varepsilon,y_\varepsilon) \leqslant f(x_\varepsilon,y)+4\varepsilon,$$

which shows that $(x_\varepsilon,y_\varepsilon)$ is an $\varepsilon$-saddle point.

Finally, for any $\varepsilon>0$ one can always find values of $x_\varepsilon$ and $y_\varepsilon$ such that inequalities (3.1) and (3.2) are satisfied (this follows directly from the definition of the extrema). Consequently, if equality (3.3) holds, we are back in the previous case [i.e. when all the three relationships (3.1), (3.2), and (3.3) are valid]. It now follows from the argument above that the function possesses a $4\varepsilon$-saddle point. To complete the proof we need only note that $\varepsilon>0$ is arbitrary. □

## 2.4 Mixed strategies

*2.4.1.* Considering examples of matrix games, we can infer that if the initially introduced (pure) strategies for the players are used, then there are antagonistic games without equilibrium situations (or even situations of $\varepsilon$-equilibrium for sufficiently small values of $\varepsilon>0$).

However, each finite (matrix) game can be formally extended to an infinite game by "supplying" the player with an arbitrary number of dominated strategies (cf. Section 1.20). Clearly, such an extension of the set of strategies for a player does not actually imply an extension of his or her possible actions and the actual behavior of a player in the extended game should not in any way differ from his or her behavior in the initial game. We have thus obtained a substantial collection of examples of infinite antagonistic games without saddle points. There are also other sources of examples of this kind.

Hence we observe that in order for the "maximin principle" to be valid in an infinite antagonistic game one should extend the class of strategic possibilities for the players. To achieve the maximin principle in the case of matrix games, it is sufficient to introduce mixed strategies for the players, i.e. probability measures on the sets of their pure strategies. Similarly, the introduction of mixed strategies for infinite antagonistic games turns out to be a fruitful idea, although it can be shown that there are infinite antagonistic games without equilibrium situations (and also without situations of $\varepsilon$-equilibrium for $\varepsilon>0$ sufficiently small) *even* in mixed strategies.

*2.4.2.* Let $\Gamma=\langle\mathfrak{X},\mathfrak{Y},H\rangle$ be an arbitrary (in general infinite) antagonistic game. As was already indicated, the mixed strategies for players in $\Gamma$ are probability distributions on the sets of their pure strategies $\mathfrak{X}$ and $\mathfrak{Y}$.

The assignment of a probability distribution on a finite set is quite a simple matter. For this purpose, it is sufficeint to assign a certain nonnegative probability to each element in such a manner that the sum of all these probabilities be equal to unity. In case of infinite sets additional care should be taken in the course of this construction: it is necessary to define a certain sufficiently wide family of subsets of the set of strategies of the player and assign to each member of this family a certain probability of choosing a pure strategy out of this particular subset. Clearly, such an assignment of probabilities must satisfy certain conditions (connected with the so-called "measurability" of the payoff function). We shall not discuss these conditions at the present time. Instead, we shall confine ourselves to a relatively elementary model of a probability distribution, assuming that it is either a finite or a countable set of points, each point having a positive measure (probability) or a *continuous* probability distribution possessing a density, or a mixture of the former and the latter.

In accordance with this simplified representation of a distribution, we shall associate with every distribution $X$ in $\mathfrak{X}$ a differential $dX(x)$. This differential represents either the probability of the "occurrence" of point $x$ under distribution $X$, or the expression $f_X(x)dx$, where $f_X(x)$ is the probability density of $X$ at point $x$, or some combination of both.

*2.4.3.* Let $X$ and $Y$ be mixed strategies for players I and II in the game $\Gamma$. The three payoffs $H(X,y)$, $H(x,Y)$, and $H(X,Y)$ are defined (by definition of the mathematical expectation) as follows:

$$H(X,y)=\int_{\mathfrak{X}}H(x,y)\,dX(x),$$

$$H(x,Y)=\int_{\mathfrak{Y}}H(x,y)\,dY(y),$$

$$\begin{aligned}H(X,Y)&=\int_{\mathfrak{X}}H(x,Y)\,dX(x)\\&=\int_{\mathfrak{Y}}H(X,y)\,dY(y)\\&=\int_{\mathfrak{X}}\int_{\mathfrak{Y}}H(x,y)\,dX(x)\,dY(y).\end{aligned}$$

*2.4.4* **Definition.** Let $\Gamma=\langle\mathfrak{X},\mathfrak{Y},H\rangle$ be an antagonistic game and let $(X^*,Y^*)$ be a situation in mixed strategies for this game. This situation is called an *equilibrium situation in mixed strategies* (or synonymously a *saddle point in mixed strategies*) if for any mixed strategies $X$ and $Y$ for players I and II, respectively, the inequality

$$H(X, Y^*) \leqslant H(X^*, Y^*) \leqslant H(X^*, Y)$$

is satisfied.

*2.4.5.* For mixed strategies in infinite antagonistic games one can state and prove assertions analogous to those stated and proved for matrix games in Section 1.10.

**Lemma** (on transition to mixed strategies). *If*

$$H(x, Y) \leqslant a \quad \text{for all } x \in \mathfrak{X}, \tag{4.1}$$

*then $H(X, Y) \leqslant a$ for all mixed strategies $X$ for player* I.

To prove the lemma, we integrate inequality (4.1) over $\mathfrak{X}$ with respect to the integrating function $X(x)$ and obtain

$$\int_{\mathfrak{X}} H(x, Y)\, dX(x) \leqslant a \int_{\mathfrak{X}} dX(x),$$

which yields the required result. □

In the same manner one can pass on to mixed strategies in the inequalities of the type

$$H(x, Y) \geqslant a \quad \text{for all } x \in \mathfrak{X},$$
$$H(X, y) \leqslant a \quad \text{for all } y \in \mathfrak{Y},$$
$$H(X, y) \geqslant a \quad \text{for all } y \in \mathfrak{Y}.$$

*2.4.6* **Theorem.** *In order that the situation $(X^*, Y^*)$ be an equilibrium situation in an antagonistic game $\Gamma = \langle \mathfrak{X}, \mathfrak{Y}, H \rangle$, it is necessary and sufficient that for all $x \in \mathfrak{X}$ and $y \in \mathfrak{Y}$ the following inequality be satisfied:*

$$H(x, Y^*) \leqslant H(X^*, Y^*) \leqslant H(X^*, y).$$

Proof. The necessity of the condition is obvious. To prove its sufficiency one needs only to pass on to mixed strategies and apply the lemma proved in Section 2.4.5 on transition to mixed strategies. □

*2.4.7* **Lemma.** *Let $\Gamma = \langle \mathfrak{X}, \mathfrak{Y}, H \rangle$ be an antagonistic game. For any mixed strategy $Y$ for player* II *we have the equality*

$$\sup_x H(x, Y) = \sup_X H(X, Y), \tag{4.2}$$

*where the supremum on the left-hand side is taken over all the pure strategies for player* I *and on the right-hand side over all his or her mixed strategies.*

*Analogously for any mixed strategy $X$ for player* I *we have the equality*

$$\inf_y H(X, y) = \inf_Y H(X, Y), \tag{4.2'}$$

*where the infimum on the left-hand side of* (4.2′) *is taken over all the pure strategies for player* II, *while on the right-hand side of* (4.2′) *over all his or her mixed strategies.*

The *proof* is presented only for the first part of the lemma.

Since the set of mixed strategies contains all the pure ones, we have

$$\sup_x H(x, Y) \leqslant \sup_X H(X, Y). \tag{4.3}$$

Assume that the strict inequality is valid in (4.3):

$$\sup_x H(x, Y) < \sup_X H(X, Y).$$

This implies that there exists $X'$ such that for some $\varepsilon > 0$

$$\sup_x H(x, Y) < H(X', Y) - \varepsilon,$$

i.e., for all $x \in \mathfrak{X}$,

$$H(x, Y) < H(X', Y) - \varepsilon.$$

Passing on in the last inequality to mixed strategies, we obtain $H(X', Y) < H(X', Y) - \varepsilon$, which is a contradiction. Hence the equality is actually attained in (4.3). □

*2.4.8* **Lemma.** *Let* $\Gamma = \langle \mathfrak{X}, \mathfrak{Y}, H \rangle$ *be an antagonistic game, let* $Y$ *be an arbitrary strategy for player* II, *and let the supremum* $\sup_X H(X, Y)$ *be attained at some (generally mixed) strategy* $X_0$. *Then the set* $\omega$ *of* $x \in \mathfrak{X}$ *such that*

$$H(x, Y) < H(X_0, Y) \tag{4.4}$$

*is realized under the distribution* $X_0$ *with probability* 0, *i.e.,* $X_0(\omega) = 0$.

*Analogously, for an arbitrary strategy* $X$ *for player I and for the strategy* $Y_0$ *at which the infimum* $\inf_Y H(X, Y)$ *is attained, the set of all* $y$ *satisfying*

$$H(X, Y_0) < H(X, y)$$

*is realized under the distribution* $Y_0$ *with probability* 0.

PROOF. Consider a sequence of positive numbers $\varepsilon_1 > \varepsilon_2 > \ldots > \varepsilon_n > \ldots > 0$ decreasing and converging to zero and denote by $\omega_n$ the set of $x$ such that the inequality

$$H(x, Y) < H(X_0, Y) - \varepsilon_n \tag{4.5}$$

is satisfied.

Clearly the sequence of sets $\omega_n$ is increasing, i.e., $\omega_1 \subset \omega_2 \subset \ldots \subset \omega_n \subset \ldots$ and $\bigcup_{n=1}^{\infty} \omega_n = \omega$.

Since the probability distribution $X_0$ is continuous, it follows that

$$\lim_{n \to \infty} X_0(\omega_n) = X_0(\omega).$$

Assume now that $X_0(\omega) > 0$, then an integer $n$ exists such that $X_0(\omega_n) > 0$.

Next, we have

$$H(X_0, Y) = \int_{\mathfrak{X}} H(x, Y)\, dX_0(x)$$
$$= \int_{\omega_n} H(x, Y)\, dX_0(x) + \int_{\mathfrak{X}\setminus\omega_n} H(x, Y)\, dX_0(x).$$

The inequality (4.5) is satisfied on the set $\omega_n$ while the inequality $H(x, Y) \leqslant H(X_0, Y)$ is valid on the complementary set $\mathfrak{X} - \omega_n$. Consequently,

$$H(X_0, Y) \leqslant \int_{\omega_n} (H(X_0, Y) - \varepsilon_n)\, dX_0(x) + \int_{\mathfrak{X}\setminus\omega_n} H(X_0, Y)\, dX_0(x)$$
$$= -\varepsilon_n \int_{\omega_n} dX_0(x) + H(X_0, Y)\left( \int_{\omega_n} dX_0(x) + \int_{\mathfrak{X}\setminus\omega_n} dX_0(x) \right)$$
$$= -\varepsilon_n \int_{\omega_n} dX_0(x) + H(X_0, Y),$$

or

$$H(X_0, Y) + \varepsilon_n X_0(\omega_n) \leqslant H(X_0, Y).$$

However, the last inequality leads to a contradiction since, by assumption, $\varepsilon_n > 0$ and $X_0(\omega_n) > 0$. Hence the only possibility is that $X_0(\omega) = 0$.

The second part of the lemma is proved analogously. □

*2.4.9* **Lemma.** *In an antagonistic game* $\Gamma = \langle \mathfrak{X}, \mathfrak{Y}, H \rangle$ *for any mixed strategy* $Y$ *for player* II *if one of the maxima*

$$\max_x H(x, Y) \quad \text{and} \quad \max_X H(X, Y) \tag{4.6}$$

*exists, so does the other and conversely; moreover, if the maxima exist they are equal.*

*Analogously for any mixed strategy* $X$ *for player* I, *if one of the minima* $\min_y H(X, y)$ *and* $\min_Y H(X, Y)$ *exists, so does the other and conversely; moreover, if these minima exist, they are equal.*

Proof. Consider equality (4.2). If the left-hand supremum is attained, so is the right-hand one, since any pure strategy is also a mixed strategy.

Assume that the right-hand supremum is attained in (4.2). Then the conditions of the lemma in 2.4.8 are fulfilled. In view of this lemma the set of $x$, such that the equality

$$H(x, Y) = \max_X H(X, Y)$$

is satisfied, is of the "full" measure (i.e., of probability 1) and is therefore nonempty. Consequently, the left-hand supremum is also attained.

Hence the maxima (4.6) exist or do not exist simultaneously. If they exist, they are equal to the corresponding suprema. In view of the lemma in Section 2.4.7 they are equal to each other.

The second part of the lemma is proved analogously. □

## 2.5 Properties of the value of a game and of optimal strategies

*2.5.1.* The assertions presented in this section repeat essentially analogous statements concerning matrix games to be found in Sections 1.15 and 1.18.

**Definition.** If equality

$$\sup_X \inf_Y H(S,Y) = \inf_Y \sup_X H(X,Y)$$

is satisfied, then the common value of these mixed extrema is called the *value* of the game with the payoff function $H$.

It follows from the theorem in Section 2.3 that a necessary and sufficient condition for a game to have a value is that for any $\varepsilon > 0$ the game possesses an $\varepsilon$-saddle point (in mixed strategies).

*2.5.2* **Theorem.** *For any antagonistic game $\Gamma = \langle \mathfrak{X}, \mathfrak{Y}, H \rangle$ that has the value $v(\Gamma)$, the following inequality is satisfied:*

$$\sup_x \inf_y H(x,y) \leqslant v(\Gamma) \leqslant \inf_y \sup_x H(x,y). \tag{5.1}$$

PROOF. In view of the lemma in Section 2.4.7, we have

$$\inf_y H(X,y) = \inf_Y H(X,Y),$$

where $X$ is an arbitrary strategy for player I. Therefore

$$\sup_x \inf_y H(x,y) \leqslant \sup_X \inf_y H(X,y) = \sup_X \inf_Y H(X,Y) = v(\Gamma),$$

and the left-hand side of (5.1) is verified. The right-hand side of this inequality is proved analogously. □

*2.5.3* **Theorem.** *If in an antagonistic game $\Gamma = \langle \mathfrak{X}, \mathfrak{Y}, H \rangle$ player* I *possesses a pure optimal strategy $x_0$ and player* II *an arbitrary (in general mixed) optimal strategy, then*

$$v(\Gamma) = \max_x \min_y H(x,y) = \min_y H(x_0,y). \tag{5.2}$$

*Analogously, if player* II *possesses a pure optimal strategy $y_0$ and player* I *an arbitrary optimal strategy, then*

$$v(\Gamma) = \min_y \sup H(x,y) = \max_x H(x,y_0).$$

PROOF. We shall carry out the proof for the first part of the theorem.

Let $Y_0$ be an optimal strategy for player II. We have from the definition of optimal strategies

$$v(\Gamma) = H(x_0, Y_0) = \min_Y H(x_0, Y).$$

Utilizing the lemma in Section 2.4.9 (concerning the existence and equality of the corresponding extrema of a payoff function) and the preceding theorem, we obtain

$$v(\Gamma)=\min_y H(x_0,y)=\inf_y H(x_0,y)\leqslant \sup_x \inf_y H(x,y)\leqslant v(\Gamma), \tag{5.3}$$

so that

$$v(\Gamma)=\inf_y H(x_0,y)=\sup_x \inf_y H(x,y).$$

Thus, the outer extrema appearing on the right-hand side of the last equality is actually attained (namely for $x=x_0$) and

$$v(\Gamma)=\max_x \inf_y H(x,y). \tag{5.4}$$

Equations (5.3) and (5.4) yield (5.2). □

*2.5.4* **Theorem.** *Let $v(\Gamma)$ be the value of the game $\Gamma=\langle\mathfrak{X},\mathfrak{Y},H\rangle$ and $v$ be an arbitrary number. If $X_0$ is a strategy for player* I, *then the inequality*

$$v\leqslant H(X_0,y)\quad \text{for all } y\in\mathfrak{Y} \tag{5.5}$$

*implies*

$$v\leqslant v(\Gamma). \tag{5.6}$$

*If $Y_0$ is a strategy for player* II, *then*

$$H(x,Y_0)\leqslant v\quad \text{for all } x\in\mathfrak{X} \tag{5.5'}$$

*implies*

$$v(\Gamma)\leqslant v. \tag{5.6'}$$

PROOF. We shall prove the first part of the theorem. Since (5.5) is valid for any $y\in\mathfrak{Y}$, passing on to mixed strategies we obtain for an arbitrary mixed strategy $Y$ for player II

$$v\leqslant H(X_0,Y).$$

Consequently, $v\leqslant\inf_Y H(X,Y)$ and thus

$$v\leqslant \sup_X \inf_Y H(X,Y). \tag{5.7}$$

However, in view of the definition in Section 2.5.1

$$\sup_X \inf_Y H(X,Y)=v(\Gamma). \tag{5.8}$$

Inequalities (5.7) and (5.8) now yield (5.6). □

*2.5.5* **Theorem.** *Let $v(\Gamma)$ be the value of the game $\Gamma=\langle\mathfrak{X},\mathfrak{Y},H\rangle$. In order that strategy $X^*$ for player* I *be optimal, it is necessary and sufficient that*

$$H(X^*,y)\geqslant v(\Gamma)\quad \text{for all } y\in\mathfrak{Y}. \tag{5.9}$$

*In order that strategy $Y^*$ for player* II *be optimal, it is necessary and* *sufficient that*

$$H(x, Y^*) \leqslant v(\Gamma) \quad \text{for all } x \in \mathfrak{X}.$$

Only the first part of the theorem is proved.

*Necessity.* The optimality of $X^*$ implies

$$\inf_Y H(X^*, Y) = \sup_X \inf_Y H(X, Y) = v(\Gamma),$$

and hence (5.9) is satisfied.

*Sufficiency.* Since (5.9) is valid for all $y \in \mathfrak{Y}$, passing on to mixed strategies we obtain for any mixed strategy $Y$ for player II

$$H(X^*, Y) \geqslant v(\Gamma)$$

and hence

$$\inf_Y H(X^*, Y) \geqslant v(\Gamma) = \sup_X \inf_Y H(X, Y),$$

i.e. the maximum of the function $\inf_Y H(X, Y)$ is attained at $X^*$. In view of the result presented in Section 1.7, this implies that $X^*$ is optimal. □

## 2.6 The Helly metric

*2.6.1.* The actual differences between two strategies in a game are due not to the outward appearances of the actions implied by these strategies but to the *consequences of using these strategies*, i.e., to the differences in the corresponding payoffs for the player who uses these strategies. This fact suggests that one should introduce the following function defined on the set of pairs of strategies for player I in an antagonistic game $\Gamma = \langle \mathfrak{X}, \mathfrak{Y}, H \rangle$:

$$\rho(x_1, x_2) = \sup_{y \in \mathfrak{Y}} |H(x_1, y) - H(x_2, y)|. \tag{6.1}$$

It is easy to verify that this function is symmetric:

$$\begin{aligned} \rho(x_1, x_2) &= \sup_y |H(x_1, y) - H(x_2, y)| \\ &= \sup_y |H(x_2, y) - H(x_1, y)| \\ &= \rho(x_2, x_1). \end{aligned}$$

Moreover, the function satisfies the triangular inequality. Indeed, for any three strategies $x_1$, $x_2$, and $x_3$, we have

$$\begin{aligned} \rho(x_1, x_2) + \rho(x_2, x_3) &= \sup_y |H(x_1, y) - H(x_2, y)| + \sup_y |H(x_2, y) - H(x_3, y)| \\ &\geqslant \sup_y (|H(x_1, y) - H(x_2, y)| + |H(x_2, y) - H(x_3, y)|) \\ &\geqslant \sup_y |H(x_1, y) - H(x_2, y) + H(x_2, y) - H(x_3, y)| \\ &\geqslant \sup_y |H(x_1, y) - H(x_3, y)| = \rho(x_1, x_3). \end{aligned}$$

Finally, it is obvious that the function is nonnegative, i.e., $\rho(x_1,x_2) \geqslant 0$ for all $x_1$ and $x_2$.

It should be pointed out, however, that the equality $\rho(x_1,x_2)=0$ does not imply that $x_1$ and $x_2$ are the *same* strategies. However, it implies that $H(x_1,y)=H(x_2,y)$ for all $y \in \mathfrak{Y}$, i.e., that the *consequences* of choosing strategies $x_1$ and $x_2$ in all situations are the same. Thus there is actually no difference between the strategies $x_1$ and $x_2$, and they can be visualized as two copies of the very same strategy. If we initially identify two strategies $x_1$ and $x_2$ for which $\rho(x_1,x_2)=0$ as the same strategy, then the condition $\rho(x_1,x_2)=0$ will imply that $x_1=x_2$. Thus under this preliminary identification the function $\rho$ satisfies all the conditions of a distance, and it can be considered as a *metric* on the space of strategies for player I.

*2.6.2* **Definition.** The metric on the space of strategies for player I defined by (6.1)—under the stipulation that $\rho(x_1,x_2)=0$ implies $x_1=x_2$—is called the *natural metric* on the space of strategies for player I or the *Helly metric*.

The topology generated by the natural metric is called *the natural topology*.

Analogously, one can introduce Helly's metric on the space of all the strategies for player II. To do this it is sufficient to define

$$\rho(y_1,y_2)=\sup_{x \in \mathfrak{X}} |H(x,y_1)-H(x,y_2)| \tag{6.1'}$$

and to stipulate that $\rho(y_1,y_2)=0$ implies $y_1=y_2$.

## 2.7 Conditionally compact games*

*2.7.1* **Definition.** A metric space $P$ is called *conditionally compact* if for any $\varepsilon>0$ there exists a *finite* $\varepsilon$-net in $P$. [Recall that the set $\mathfrak{X}_\varepsilon$ is called an $\varepsilon$-net in the space $\mathfrak{X}$ with metric $\rho$ if for any $x \in \mathfrak{X}$ there exists $x_\varepsilon \in \mathfrak{X}_\varepsilon$ such that $\rho(x,x_\varepsilon)<\varepsilon$.]

Conditionally compact spaces are also often called *completely bounded* (or *subcompact* in more recent terminology).

Conditionally compact spaces are widely used in various branches of mathematical analysis. For example, any bounded subset of a finite-dimensional Euclidean space is conditionally compact in the Euclidean metric.

*2.7.2* **Definition.** An antagonistic game $\Gamma=\langle \mathfrak{X},\mathfrak{Y},H \rangle$ is called *conditionally compact* if the spaces of strategies for both players in the Helly metric are conditionally compact.

It turns out that in order for a game to be conditionally compact, it is

*This section requires knowledge of some advanced mathematical notions and may be omitted in the first reading (Translator's remark).

sufficient to require the conditional compactness in the Helly metric of the space of strategies for one of the players *only*.

*2.7.3* **Theorem.** *If in a game $\Gamma=\langle \mathfrak{X},\mathfrak{Y},H\rangle$ the space of strategies for one of the players is conditionally compact in the Helly metric, so then is the space of strategies of the second player in his or her Helly's metric.*

PROOF. Let $\mathfrak{X}$ be conditionally compact in the Helly metric. Choose an arbitrary $\varepsilon>0$ and select finite $\varepsilon/3$-net $x_1,\ldots,x_m$ in $\mathfrak{X}$. Now correspond to each $y\in\mathfrak{Y}$ the $m$-dimensional vector

$$\phi(y)=(H(x_1,y),\ldots,H(x_m,y)).$$

The set of all of these vectors forms a bounded subset of the $m$-dimensional Euclidean space and is therefore conditionally compact in the Euclidean metric $\rho_E$. Consequently, there exists in this space a finite $\varepsilon/3$-net in metric $\rho_E$. Let

$$y_1,\ldots,y_m \tag{7.1}$$

be the strategies for player II corresponding to the vectors that form this (finite) net. We now show that they constitute an $\varepsilon$-net in space $\mathfrak{Y}$ in the Helly metric.

Choose an arbitrary strategy $y$ for player II. According to the condition there exists $y_j$ belonging to the finite set $y_1,\ldots,y_m$ given by (7.1) such that

$$\rho_E(\phi(y),\phi(y_j))<\varepsilon/3.$$

However,

$$\rho_E(\phi(y),\phi(y_j))=\sqrt{\sum_{i=1}^{m}(H(x_i,y)-H(x_i,y_j))^2}$$
$$\geqslant \sup_i|H(x_i,y)-H(x_i,y_j)|.$$

Hence also

$$\sup_i|H(x_i,y)-H(x_i,y_j)|>\varepsilon/3.$$

Correspond to each $x\in\mathfrak{X}$ an $x_i$ belonging to the $\varepsilon/3$-net such that $\rho(x,x_i)<\varepsilon/3$ and denote this $x_i$ by $x_\varepsilon$. We have

$$\begin{aligned}
\rho(y,y_j)&=\sup_x|H(x,y)-H(x,y_j)|\\
&=\sup_x|H(x,y)-H(x_\varepsilon,y)+H(x_\varepsilon,y)-H(x_\varepsilon,y_j)+H(x_\varepsilon,y_j)-H(x,y_j)|\\
&\leqslant \sup_x|H(x,y)-H(x_\varepsilon,y)|+\sup_x|H(x_\varepsilon,y)-H(x_\varepsilon,y_j)|\\
&\quad+\sup_x|H(x_\varepsilon,y_j)-H(x,y_j)|\\
&\leqslant \varepsilon/3+\varepsilon/3+\varepsilon/3=\varepsilon.
\end{aligned}$$

Thus for any $\varepsilon>0$ a finite $\varepsilon$-net has been constructed in the space $\mathfrak{Y}$. This shows that the space $\mathfrak{Y}$ is conditionally compact. □

## 2.8 The basic theorem for conditionally compact games*

*2.8.1* **Theorem.** *If an antagonistic game $\Gamma=\langle\mathfrak{X},\mathfrak{Y},H\rangle$ is conditionally compact, then it possesses $\varepsilon$-optimal mixed strategies for any $\varepsilon>0$.*

PROOF. Choose an arbitrary $\varepsilon>0$ and form the two $\varepsilon$-nets

$$\mathfrak{X}_\varepsilon : x_1,\dots,x_m, \qquad \mathfrak{Y}_\varepsilon : y_1,\dots,y_n$$

in the corresponding Helly metrics of the spaces $\mathfrak{X}$ and $\mathfrak{Y}$. Consider the antagonistic game

$$\Gamma_\varepsilon=\langle\mathfrak{X}_\varepsilon,\mathfrak{Y}_\varepsilon,H_\varepsilon\rangle,$$

where for any $x\in\mathfrak{X}_\varepsilon$ and $y\in\mathfrak{Y}_\varepsilon$

$$H_\varepsilon(x,y)=H(x,y).$$

Clearly, one can replace $H_\varepsilon$ by $H$ everywhere without causing any ambiguity.

The game $\Gamma_\varepsilon$ is a finite game, namely, a matrix game, and as such it possesses equilibrium situations in mixed strategies. Let $(X_\varepsilon, Y_\varepsilon)$ be one of these equilibrium situations.

We have $X_\varepsilon=(\xi,\dots,\xi_m)$ and $Y_\varepsilon=(\eta_1,\dots,\eta_n)$, where $\xi_i$ is the probability of choosing strategy $x_i\in\mathfrak{X}_\varepsilon$ and $\eta_j$ is the corresponding probability of choosing strategy $y_i\in\mathfrak{Y}_\varepsilon$.

The equilibrium property of situation $(X_\varepsilon, Y_\varepsilon)$ in the game $\Gamma_\varepsilon$ implies that

$$H(x_i,Y_\varepsilon)\leqslant H(X_\varepsilon,Y_\varepsilon)\leqslant H(X_\varepsilon,y_j) \quad \text{for all } x_i\in\mathfrak{X}_\varepsilon \text{ and } y_j\in\mathfrak{Y}_\varepsilon. \tag{8.1}$$

Given an arbitrary $x\in\mathfrak{X}$ we have $\rho(x,x_\varepsilon)<\varepsilon$, where $x_\varepsilon$ is the corresponding "neighboring" strategy $x_\varepsilon\in\mathfrak{X}_\varepsilon$; in other words

$$\sup_y|H(x,y)-H(x_\varepsilon,y)|<\varepsilon,$$

and in particular for all $j=1,\dots,n$,

$$|H(x,y_j)-H(x_\varepsilon,y_j)|<\varepsilon;$$

whence

$$H(x,y_j)-\varepsilon\leqslant H(x_i,y_j), \qquad i=1,\dots,m, \quad j=1,\dots,n.$$

Therefore,

$$\begin{aligned}
H(x_i,Y_\varepsilon)&=\sum_{j=1}^{n}H(x_i,y_j)\eta_j\\
&\geqslant\sum_{j=1}^{n}(H(x,y_j)-\varepsilon)\eta_j\\
&=\sum_{j=1}^{n}H(x,y_j)\eta_j-\varepsilon\\
&=H(x,Y_\varepsilon)-\varepsilon.
\end{aligned}$$

*This section requires knowledge of some advanced mathematical notions and may be omitted in the first reading (Translator's remark).

In the same manner one shows that for any strategy $y$ and its "neighboring" strategy $y_\varepsilon = y_j \in \mathfrak{Y}_\varepsilon$

$$H(X_\varepsilon, y_j) \leqslant H(X_\varepsilon, y) + \varepsilon.$$

Substituting the bounds obtained for $H(x_i, Y_\varepsilon)$ and $H(X_\varepsilon, y_j)$ into (8.1), we obtain

$$H(x, Y_\varepsilon) - \varepsilon \leqslant H(X_\varepsilon, Y_\varepsilon) \leqslant H(X_\varepsilon, y) + \varepsilon \quad \text{for all } x \in \mathfrak{X}, \text{ and } y \in \mathfrak{Y}.$$

The theorem is thus proved. □

## 2.9 Continuous games on the unit square

*2.9.1* **Definition.** An antagonistic game $\Gamma = \langle \mathfrak{X}, \mathfrak{Y}, H \rangle$ is called a *game on the unit square* if the sets of pure strategies for each of the players are the unit segments

$$\mathfrak{X} = \mathfrak{Y} = [0, 1].$$

The terminology is motivated by the fact that the set of all situations in this game is the Cartesian product of two closed unit segments $[0, 1]$ forming the unit square.

*2.9.2.* Mixed strategies for the players in a game on the unit square are probability distributions on unit segments.

For this type of mixed strategies the following theorems are valid.*

**Helly's first theorem.** *For any sequence of mixed strategies for one of the players (i.e. probability distributions on a segment) one can select a subsequence that is weakly convergent (i.e. that is convergent at each point of continuity of the limiting distribution function).*

**Helly's second theorem.** *For any sequence of mixed strategies for a player (i.e. distribution functions on a segment) $Z_1, \ldots, Z_n, \ldots$ that is weakly convergent to the limit $Z_0$ and for any continuous function $f(z)$ in the variable $z$, the following equality is valid:*

$$\lim_{n \to \infty} \int_{\mathfrak{Z}} f(z)\, dZ_n(z) = \int_{\mathfrak{Z}} f(z)\, dZ_0(z).$$

*In particular, if the sequence of mixed strategies for player* I, $X_1, \ldots, X_n, \ldots,$ *is weakly convergent to $X_0$ and the payoff function $H(x, Y)$ is continuous in $x$, then*

$$\lim_{n \to \infty} H(X_n, Y) = H(X_0, Y).$$

* See, e.g., Gnedenko, B. V. *The Theory of Probability*, Chelsea Publ. Co., N.Y. (1962), 263–268, or W. Feller, *An Introduction to Probability Theory and Its Applications*, Vol. II, 2nd Edition, John Wiley and Sons, Inc., New York (1971), 267–268.

*Analogously, if a sequence of mixed strategies for player* II *is weakly convergent to* $Y_0$ *and the payoff function* $H(X,y)$ *is continuous in* $y$*, then*

$$\lim_{n\to\infty} H(X, Y_n) = H(X, Y_0).$$

*2.9.3* **Definition.** A game on the unit square is called *continuous* if the payoff function $H$ is continuous in both (numerical) variables.

**Lemma.** *If the function* $H(x,y)$ *is continuous on the unit square in both variables* $x$ *and* $y$*, and* $X$ *is an arbitrary probability measure on* $[0,1]$*, then the integral*

$$H(X,y) = \int_0^1 H(x,y)\,dX(x)$$

*is also a continuous function in* $y$*.*
*Analogously, for any probability measure* $Y$ *on* $[0,1]$ *the integral*

$$H(x,Y) = \int_0^1 H(x,y)\,dY(y)$$

*is also a continuous function in* $x$*.*

Proof. Since the unit square is a closed set, a continuous function defined on it is uniformly continuous. Choose an arbitrary $\varepsilon > 0$; in view of the uniform continuity of $H$ there exists $\delta$ such that

$$|H(x,y_1) - H(x,y_2)| < \varepsilon$$

for all $x$, $y_1$, and $y_2$ provided only that $|y_1 - y_2| < \delta$.
This implies that

$$\begin{aligned} |H(X,y_1) - H(X,y_2)| &= \left| \int_0^1 H(x,y_1)\,dX(x) - \int_0^1 H(x,y_2)\,dX(x) \right| \\ &= \left| \int_0^1 (H(x,y_1) - H(x,y_2))\,dX(x) \right| \\ &\leqslant \int_0^1 |H(x,y_1) - H(x,y_2)|\,dX(x) \\ &< \int_0^1 \varepsilon\,dX(x) = \varepsilon. \end{aligned}$$

Consequently the function $H(X,y)$ is continuous in $y$. □

*2.9.4* **Theorem** (The basic theorem for continuous games on the unit square). *For any continuous game* $\Gamma$ *on the unit square with the payoff function* $H$ *both players possess optimal mixed strategies.*

Proof. Since the payoff function $H$ is continuous on the closed unit square, it is uniformly continuous on this square. Therefore for any $\varepsilon > 0$

one can find a $\delta > 0$ such that for any $y \in [0, 1]$

$$|H(x_1,y) - H(x_2,y)| < \varepsilon/2,$$

provided only $|x_1 - x_2| < \delta$.

Consequently,

$$\sup|H(x_1,y) - H(x_2,y)| \leqslant \varepsilon/2 < \varepsilon,$$

i.e. $\rho(x_1, x_2) < \varepsilon$. This implies that the correspondence between $\varepsilon$ and $\delta$ is such that a $\delta$-net (in the usual Euclidean metric) on the segment $[0, 1]$ is an $\varepsilon$-net on this segment in the sense of the Helly metric.

Thus for any $\varepsilon > 0$ there exists a finite $\varepsilon$-net on the space of strategies for player I. Consequently, the game is conditionally compact and therefore, in view of the theorem on conditionally compact games (Section 2.8), the game possesses situations of $\varepsilon$-equilibrium for any $\varepsilon > 0$. Therefore the value is defined for this game.

Denote the value of game $\Gamma$ by $v$ and choose a decreasing sequence $\{\varepsilon_n\}$

$$\varepsilon_1 > \varepsilon_2 > \ldots > \varepsilon_n > \ldots > 0$$

convergent to zero. For each $n = 1, 2, \ldots$ we construct an $\varepsilon_n$-saddle point $(X_n, Y_n)$.

In view of the first assertion of the theorem in Section 2.3 we have

$$\inf_Y H(X_n, Y) + 2\varepsilon_n \geqslant v,$$

so that for any pure strategy $y$ for player II

$$H(X_n, y) + 2\varepsilon_n \geqslant v. \tag{9.1}$$

The first Helly's theorem (cf. Section 2.9.2) yields that one can select from the sequence

$$X_1, X_2, \ldots, X_n, \ldots \tag{9.2}$$

a subsequence that is weakly convergent to a certain strategy (i.e. distribution function) $X_0$. Without loss of generality it may be assumed that (9.2) is the required subsequence.

We now approach the limit in (9.1) and obtain

$$\lim_{n\to\infty} H(X_n, y) + \lim_{n\to\infty} 2\varepsilon_n \geqslant v.$$

In view of Helly's second theorem (cf. Section 2.9.2) the first of these limits is $H(X_0, y)$ and the second is trivially zero. Thus we have $H(X_0, y) \geqslant v$ for any $y$. Consequently, in view of the theorem in Section 2.5.5, $X_0$ is an optimal strategy for player I in the game under consideration. A symmetric argument proves the existence of an optimal strategy for player II. □

*2.9.5* **Definition.** Let $\Gamma$ be a game on the unit square. A pure strategy $z$ for a player is called a *point of the spectrum* of the mixed strategy $Z$ if for any neighborhood $\omega$ of point $z$ the value of the integral $\int_\omega dZ(z)$ is positive:

$$\int_\omega dZ(z) > 0.$$

*2.9.6* **Theorem.** *Let* $\Gamma$ *be a continuous game on the unit square with the payoff function* $H$ *and value* $v$.

*If* $Y^*$ *is an arbitrary optimal strategy for player* II *and if for some* $x_0$

$$H(x_0, Y^*) < v, \tag{9.3}$$

*then* $x_0$ *is not a point of the spectrum of an optimal strategy for player* I.

*If* $X^*$ *is an arbitrary optimal strategy for player* I *and if for some* $y_0$

$$H(X^*, y_0) > v, \tag{9.3'}$$

*then* $y$ *is not a point of the spectrum of an optimal strategy for player* II.

PROOF. We shall prove the first part of the theorem only.

Let $X^*$ be an optimal strategy for player I. Then

$$H(X^*, Y^*) = v = \max_X H(X, Y^*).$$

Consequently, in view of the lemma in Section 2.4.8, the set $\omega$ of all $x$, such that

$$H(x, Y^*) < v, \tag{9.4}$$

satisfies

$$X^*(\omega) = 0. \tag{9.5}$$

Furthermore, in view of the lemma in Section 2.9.3, the function $H(x, Y^*)$ is continuous. Therefore, (9.3) implies that there exists a neighborhood $\omega_0$ of the point $x_0$ such that for all $x \in \omega_0$ (9.4) is satisfied. Since $\omega_0 \subset \omega$, (9.5) yields that

$$X^*(\omega_0) = 0. \tag{9.6}$$

On the other hand, if $x_0$ were a point of the spectrum, we would have $X^*(\omega_0) > 0$. This contradiction proves the theorem. □

## 2.10 Convex functions

*2.10.1.* So far no general methods for determining solutions of infinite antagonistic games including continuous games on the unit square have been discovered. Only special individual tricks have been developed which are applicable to a rather narrow class of games of this type. One such class is the class of games with a convex payoff function. These games are also of some interest in practical applications.

**Definition.** A function $\phi$ defined on the segment $[a, b]$ of the real line is called *convex* if for any $x_1, x_2 \in [a, b]$ and for any $\lambda \in [0, 1]$ the inequality

$$\phi(\lambda x_1 + (1-\lambda)x_2) \leqslant \lambda\phi(x_1) + (1-\lambda)\phi(x_2) \tag{10.1}$$

is satisfied.

If for all $x_1 \neq x_2$ and for all $\lambda$ (except $\lambda \neq 0$ and $\lambda \neq 1$) the strict inequality is attained in (10.1) then the function $\phi$ is called *strictly convex*.

In what follows we shall restrict our attention to the case of the *continuous convex function*.*

The geometric convexity of a function corresponds to its graph being convex *downward*. In other words, each point of the chord joining any two points on the graph of this function is above the corresponding graph. (See Figure 16.)

Analytically, if a convex function is twice differentiable, then its second derivative is nonnegative (and is positive for a strictly convex function).

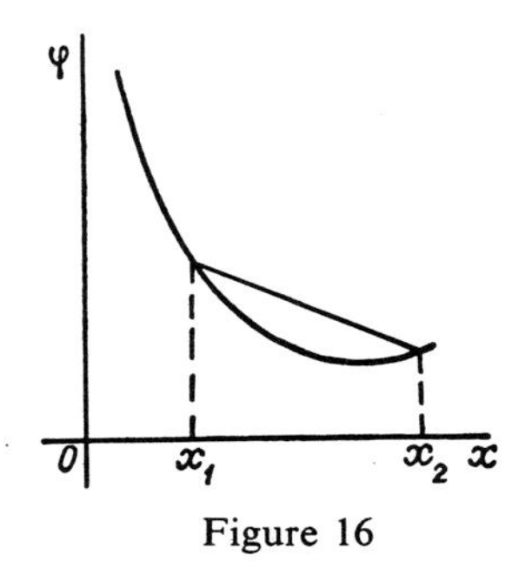


Figure 16

*2.10.2* **Lemma** (discrete form of Jensen's Inequality). *If $\phi$ is a convex function on $[a,b]$,*

$$x_1,\ldots,x_n \in [a,b],$$

*and $\lambda_1,\ldots,\lambda_n$ is a sequence of nonnegative numbers whose sum is 1, then*

$$\phi\left(\sum_{i=1}^{n} \lambda_i x_i\right) \leqslant \sum_{i=1}^{n} \lambda_i \phi(x_i).$$

This assertion is proved directly by inductive application of the definition of convexity of a function.

*2.10.3* **Lemma** (Jensen's Inequality). *If a function is convex on a closed interval $[a,b]$ and $X$ is an arbitrary probability measure on $[a,b]$, then*

$$\phi\left(\int_a^b x\,dX(x)\right) \leqslant \int_a^b \phi(x)\,dX(x).$$

The proof follows from the lemma in the previous item by approaching the limit and taking into account the continuity of the function†.

* It should be noted that any convex function defined on an *open* interval $(a,b)$ is absolutely continuous on each *closed* subinterval of $(a,b)$. (See e.g., Royden, H. L. *Real Analysis*, 2nd Edition, Macmillan, New York (1961), 109) (Translator's remark).

† A detailed proof is given, for example, in W. Rudin *Real and Complex Analysis*, McGraw-Hill, New York (1966), 61 (Translator's remark).

*2.10.4* **Lemma.** *If a function $\phi(x,y)$ is convex in $y$ for any $x\in[a,b]$ and $X(\cdot)$ is an arbitrary probability measure on $[a,b]$ then the integral*

$$\psi(y)=\int_a^b \phi(x,y)\,dX(x)$$

*is also a convex function of $y$.*

PROOF. Choose an arbitrary $\lambda\in[0,1]$. Since $\phi$ is convex, we have for any two values $y_1$ and $y_2$

$$\begin{aligned}\psi(\lambda y_1+(1-\lambda)y_2)&=\int_a^b \phi(x,\lambda y_1+(1-\lambda)y_2)\,dX(x)\\ &\leqslant\int_a^b (\lambda\phi(x,y_1)+(1-\lambda)\phi(x,y_2))\,dX(x)\\ &=\lambda\int_a^b \phi(x,y_1)dX(x)+(1-\lambda)\int_a^b \phi(x,y_2)dX(x)\\ &=\lambda\psi(y_1)+(1-\lambda)\psi(y_2),\end{aligned}\tag{10.2}$$

which proves the lemma. □

Observe that in view of the lemma in Section 2.9.3, continuity of $\phi$ implies continuity of $\psi$. Therefore the integral transformations applied in (10.2) under the conditions of the lemma do not contradict the assumption that all convex functions discussed herein are continuous.

*2.10.5* **Lemma.** *If the function $\phi(x,y)$ is strictly convex in $y$ for any $x\in[a,b]$ and $X(\cdot)$ is an arbitrary probability measure on $[a,b]$, then the integral*

$$\psi(y)=\int_a^b \phi(x,y)\,dX(x)$$

*is also a strictly convex function.*

The proof of this lemma is analogous to that of the preceding one. The strict inequality in (10.2) is assured by the fact that

$$\phi(x,\lambda y_1+(1-\lambda)y_2)<\lambda\phi(x,y_1)+(1-\lambda)\phi(x,y_2)$$

for all the values of $x$ and the set of all values of $x$ is of a positive measure with respect to the probability measure $X$.

*2.10.6* **Lemma.** *If the argument of a convex function is increased, the function cannot change its behavior from an increasing function to a decreasing one.*

PROOF. Assume that $x_1<x_2<x_3$, and let

$$\phi(x_1)<\phi(x_2)>\phi(x_3).\tag{10.3}$$

Choose $\lambda\in[0,1]$ such that $x_2=\lambda x_1+(1-\lambda)x_3$. Then since the function $\phi$ is convex, we have

$$\begin{aligned}\phi(x_2)=\phi(\lambda x_1+(1-\lambda)x_3) &\leqslant \lambda\phi(x_1)+(1-\lambda)\phi(x_3)\\ &\leqslant \max\{\phi(x_1),\phi(x_3)\},\end{aligned}$$

which contradicts (10.3). □

*2.10.7* **Corollaries.** (1) *A convex function consists at most of two monotonic parts; if there are two parts, then the decreasing part precedes the increasing one.*

(2) *The set of points at which a convex function attains its minimal values is connected* (*i.e. is a segment*).

Indeed, the fact that the set of minimal values of a convex function is nonvoid and closed follows from the assumption that the function is continuous. If between two points corresponding to the minima of the function there is a point that is not a minimum, then the function increases to the left of this point while it decreases to the right. This, however, contradicts the first corollary.

(3) *A strictly convex function attains its minimum value at exactly one point* (*i.e. it possesses a unique minimum*).

## 2.11 Convex games; pure optimal strategies for player II

*2.11.1* **Definition.** A continuous antagonistic game on the unit square with the payoff function $H$ is called *convex* if $H(x,y)$ is convex in $y$ for any value of $x$.

As is the case for continuous games on the unit square, the players in a convex game possess optimal strategies. In this section and in the next two sections, we shall describe the structure of solutions and methods of determining optimal strategies for players in convex games on the unit square.

*2.11.2* **Theorem.** *In a convex game on the unit square player* II *possesses a pure optimal strategy.*

PROOF. Consider a convex game $\Gamma$ with the payoff function $H$. Since the game is continuous, according to the basic theorem in Section 2.9.4, it possesses a value $v$ and optimal strategies for the players. Let $X^*$ be a strategy for player I and $Y^*$ an optimal strategy for player II. Then

$$H(x,Y^*)\leqslant H(X^*,Y^*)=v \quad \text{for all } x\in[0,1]. \tag{11.1}$$

Set

$$y^*=\int_0^1 y\,dY^*(x). \tag{11.2}$$

Here $y^*\in[0,1]$ and therefore is a pure strategy for player II. In view of the lemma in Section 2.10.3, we have for any $x$

$$H(x,y^*)=H\left(x,\int_0^1 y\,dY^*(x)\right)\leqslant\int_0^1 H(x,y)\,dY^*(x)=H(x,Y^*). \quad (11.3)$$

This together with (11.1) implies that $H(x,y^*)\leqslant v$ for all $x\in[0,1]$. Consequently, according to the theorem in Section 2.5.5, $y^*$ is an optimal strategy for player II. □

*2.11.3* **Corollary.** *For any convex game we have*

$$v=\min_y \max_x H(x,y)=\max_x H(x,y^*), \quad (11.4)$$

*where $y^*$ is an optimal pure strategy for player* II.

This follows from the theorem in Section 2.5.3. Moreover, the maximum with respect to $x$ in (11.4) is actually attained since the function $H$ is continuous in $x$.

*2.11.4* **Corollary.** *The pure strategies $y^*$ for player* II *in a convex game are the solutions of the equation*

$$v=\max_x H(x,y). \quad (11.5)$$

Indeed, in order that the strategy $y^*$ for player II be the solution of equation (11.5), it is necessary and sufficient that $H(x,y^*)\leqslant v$ for all $x$. However, according to the theorem in Section 2.5.5, this is equivalent to the optimality of this strategy for player II. □

## 2.12 Convex games; optimal strategies for player I

*2.12.1.* We now proceed to describe optimal strategies for player I in convex games.

We shall denote by $H_y'(x,y)$ the partial derivative of the payoff function with respect to $y$. For $y=0$ this expression is interpreted as the right-hand derivative and for $y=1$ as the left-hand one. We shall assume that $H_y'(x,y)$ exists for all the values of $x$ and $y$.

Let $y^*$ be an optimal strategy for player II. In view of the theorem in Section 2.9.6, the spectrum of optimal strategies for player I contains only those pure strategies $x$ which satisfy

$$H(x,y^*)=v. \quad (12.1)$$

Pure strategies for player I that satisfy this equality are sometimes called *essential*.

*2.12.2* **Lemma.** *If $y^*$ is an optimal strategy for player* II *in a convex game with the payoff function $H$ differentiable with respect to $y$, and if $y^*>0$, then an essential strategy $x'$ for player* I *exists such that*

$$H_y(x',y^*)\leqslant 0. \quad (12.2)$$

PROOF. Assume that for any essential strategy $x$ for player I

$$H'_y(x,y^*) > 0.$$

This implies that for any essential strategy $x$ the function $H(x,y)$ is an increasing function of $y$ at the point $y^*$. Hence for the values $y$ smaller than $y^*$ and sufficiently close to $y^*$ we have

$$H(x,y) < H(x,y^*). \tag{12.3}$$

Let $X^*$ be an optimal strategy for player I. Since, by assumption, (12.3) is valid for all essential strategies for player I and in particular for all points of the spectrum of $X^*$, integrating with respect to the distribution $X^*$ we obtain

$$H(X^*,y) < H(X^*,y^*) = v.$$

This, however, contradicts the optimality of strategy $X^*$.
The lemma is thus proved. □

*2.12.3.* The following symmetric assertion is proved analogously.

**Lemma.** *If under the conditions of the preceding lemma $y^* < 1$, an essential strategy $x''$ for player* I *exists such that*

$$H'_y(x'',y^*) \geqslant 0.$$

*2.12.4.* Combining these lemmas we obtain:

**Corollary.** *If under the conditions of the two preceding lemmas $0 < y^* < 1$, then essential strategies $x'$ and $x''$ exist for player* I *such that*

$$H'_y(x',y^*) \leqslant 0, \tag{12.4}$$

$$H'_y(x'',y^*) \geqslant 0. \tag{12.5}$$

*2.12.5* **Theorem.** *Let $\Gamma$ be a convex game with a payoff function $H$ differentiable in $y$ for any $x$; let $y^*$ be a pure optimal strategy for player* II *and let $v$ be the value of the game.*

*Then*:

(1) *if $y^* = 1$, an optimal pure strategy $x'$ exists for player* I, *which is also essential, satisfying*

$$H'_y(x',1) \leqslant 0;$$

(2) *if $y^* = 0$, then an optimal pure strategy $x''$ exists for player* I, *which is also essential, satisfying*

$$H'_y(x'',0) \geqslant 0;$$

(3) *if $0 < y' < 1$, then among optimal strategies for player* I *a strategy exists which is a mixture of two essential strategies $x'$ and $x''$. Moreover, these*

*strategies satisfy*

$$H_y'(x',y^*) \leqslant 0, \qquad H_y'(x'',y^*) \geqslant 0.$$

*Furthermore, strategies $x'$ and $x''$ are utilized with probabilities $\alpha$ and $1-\alpha$, respectively, where $\alpha$ is determined by*

$$\alpha H_y'(x',y^*) + (1-\alpha) H_y'(x'',y^*) = 0. \tag{12.6}$$

PROOF. First let $y^*=1$. In view of the lemma in Section 2.12.2 there exists an essential strategy $x'$ for player I for which (12.2) is satisfied. Consequently, in the neighborhood of $y^*=1$ the function $H(x',y)$ is a decreasing function of $y$. However, any convex function with an increase in the values of the argument cannot change its behavior from increasing to decreasing (cf. the lemma in Section 2.10.6). Therefore, $H(x',y)$ is a decreasing function on the segment $[0,1]$ and attains its minimum at $y=1$. This implies that

$$H(x',y^*) \leqslant H(x',y) \quad \text{for all } y \in [0,1]. \tag{12.7}$$

On the other hand, since the pure strategy $x'$ for player I is essential, it follows that

$$H(x,y^*) \leqslant H(x',y^*) \quad \text{for all } x \in [0,1]. \tag{12.8}$$

Inequalities (12.7) and (12.8) imply that $(x',y^*)$ is a saddle point and the case (1) of the theorem is verified.

A symmetric argument validates the case (2) of the theorem.

We now proceed to verify the case (3). According to the Corollary in Section 2.12.4, essential strategies $x'$ and $x''$ exist satisfying (12.4) and (12.5), respectively.

Consider the function

$$f(\xi) = \xi H_y'(x',y^*) + (1-\xi) H_y'(x'',y^*).$$

Inequalities (12.4) and (12.5) imply that $f(0) \leqslant 0$ and $f(1) \geqslant 0$. Since the function $f$ is continuous (in this case a linear function) an $\alpha \in [0,1]$ exists such that $f(\alpha)=0$.

Now choose a mixed strategy for player I consisting of assigning probability $\alpha$ to strategy $x'$ and probability $1-\alpha$ to strategy $x''$. Denote this strategy by $X^*$.

In view of lemma in Section 2.10.2 the function

$$H(X^*,y) = \alpha H(x',y) + (1-\alpha) H(x'',y)$$

is a convex function in $y$. Its derivative with respect to $y$ at the point $y=y^*$ is equal to

$$H_y'(X^*,y^*) = \alpha H_y'(x',y^*) + (1-\alpha) H_y'(x'',y^*) = f(\alpha) = 0.$$

Consequently, the function $H(x^*,y)$ attains an extremum at the point $y^*$; since the function is convex this extremum must be a minimum. Thus

$$H(X^*,y^*) \leqslant H(X^*,y) \quad \text{for all } y \in [0,1]. \tag{12.9}$$

On the other hand, since the strategy $x'$ is essential,

$$H(X^*,y^*) = H(x',y^*) = v = \max_x H(x,y^*)$$
$$\geqslant H(x,y^*) \quad \text{for all } x \in [0,1]. \tag{12.10}$$

Relations (12.9) and (12.10) yield that the situation $(X^*,y^*)$ is an equilibrium situation and the strategy $X^*$ is optimal. The case (3) of the theorem is verified. □

## 2.13 Strictly convex games

*2.13.1* **Definition.** A continuous antagonistic game $\Gamma$ on the unit square is called *strictly convex* if the payoff function $H(x,y)$ is a strictly convex function in $y$ for any value of $x \in [0,1]$.

*2.13.2* **Theorem.** *In a strictly convex game* $\Gamma$, *player* II *possesses a unique optimal strategy; moreover, this strategy is pure.*

Proof. Let $X^*$ be an optimal strategy for player I and let

$$\psi(y) = H(X^*,y).$$

According to the theorem in Section 2.5.6, in order that a pure strategy $y^*$ be a point of the spectrum of some optimal strategy for player II it is necessary that equality

$$\psi(y^*) = \min \psi(y) = v$$

be satisfied.

However, the lemma in Section 2.10.5 implies that the function $\psi(y)$ is a strictly convex function, and corollary (3) in Section 2.10.7 yields that it possesses a unique minimum. Denote this minimum by $y^*$. It thus follows that the spectrum of any optimal strategy for player II consists of only the single point $y^*$. This implies that there are no other optimal strategies for player II except the pure strategy $y^*$. □

## 2.14 Examples of convex games and their solutions

*2.14.1* Example. Let a game on the unit square be defined by the payoff function

$$H(x,y) = (x-y)^2. \tag{14.1}$$

In this case,

$$\frac{\partial^2 H(x,y)}{\partial y^2} = 2 > 0,$$

hence the game with the payoff function (14.1) is strictly convex. The value of the game is obtained from formula (11.4).

$$v = \min_y \max_x (x-y)^2.$$

Let $y_0$ be a fixed strategy for player II. The expression $(x-y_0)^2$ attains its maximum at $x=1$ provided $y_0 \leqslant \frac{1}{2}$, and at $x=0$ if $y_0 \geqslant \frac{1}{2}$. In other words,

$$\max_x (x-y)^2 = \begin{cases} (1-y)^2 & \text{if} \quad y \leqslant \frac{1}{2}, \\ y^2, & \text{if} \quad y \geqslant \frac{1}{2}. \end{cases}$$

Hence,

$$v = \min\left\{ \min_{0 \leqslant y \leqslant \frac{1}{2}} (1-y)^2;\ \min_{\frac{1}{2} \leqslant y \leqslant 1} y^2 \right\}.$$

In the last expression the first of the inner minima is attained at $y=\frac{1}{2}$ and the second is also attained at the same point. The value of the corresponding function at each one of these minima is therefore $\frac{1}{4}$. Hence $v=\min[\frac{1}{4},\frac{1}{4}]=\frac{1}{4}$ and the value of the game is determined. Moreover, the unique pure optimal strategy for player II is the strategy $y^*=\frac{1}{2}$.

We now proceed to determine optimal strategies for player I. Since $0<y^*=\frac{1}{2}<1$ we are dealing here with the case (3) of the classification presented in the theorem in Section 2.12.5. We shall seek essential pure strategies for player I. Equation (12.1) in the case of payoff function (14.1) reduces to

$$\left(x-\tfrac{1}{2}\right)^2=\tfrac{1}{4}.$$

Solving this equation we obtain the following two essential strategies for player I: $x_1=0$ and $x_2=1$.

Consequently, the optimal strategy for player I is the (probabilistic) mixture of his pure strategies 0 and 1.

Differentiating the payoff function (14.1) with respect to $y$, we obtain

$$\left.\frac{\partial (x_1-y)^2}{\partial y}\right|_{y=\frac{1}{2}} = 1>0,$$

$$\left.\frac{\partial (x_2-y)^2}{\partial y}\right|_{y=\frac{1}{2}} = -1<0$$

(as it should be expected from the theoretical considerations established in Section 2.12, the values of these partial derivatives are different in sign). Equation (12.6) in our case is of the form $\alpha\cdot 1+(1-\alpha)\cdot(-1)=0$, i.e. $2\alpha-1=0$, or $\alpha=\frac{1}{2}$. Thus the optimal strategy for player I consists of choosing his pure strategies 0 and 1 each with probability $\frac{1}{2}$.

*2.14.2* Example. The payoff function of a game on the unit square is given by

$$H(x,y)=y^3-3xy+x^3.$$

Here

$$\frac{\partial^2 H(x,y)}{\partial y^2} = 6y^2 > 0, \tag{14.2}$$

so that the game with the payoff function (14.2) is strictly convex except at the point $y=0$.

In view of (11.4), we have

$$v = \min_y \max_x (y^3 - 3xy + x^3).$$

Also we have for any $y$

$$\frac{\partial H(x,y)}{\partial x} = 3x^2 - 3y.$$

In other words the payoff function decreases in $x$ for $x \leqslant \sqrt{y}$ and increases for $x \geqslant \sqrt{y}$. Consequently, for any value of $y$, the maximal value of the payoff function is attained either at $x=0$ or at $x=1$. Therefore,

$$\max_x (y^3 - 3xy + x^3) = \max\{y^3, y^3 - 3y + 1\}.$$

Clearly, for $y \leqslant \frac{1}{3}$ the maximum is attained at $x=1$, and for $y \geqslant \frac{1}{3}$ at $x=0$. Thus

$$\max_x H(x,y) = \begin{cases} y^3 - 3y + 1, & \text{if} \quad y \leqslant \frac{1}{3}, \\ y^3, & \text{if} \quad y \geqslant \frac{1}{3}. \end{cases}$$

Hence

$$v = \min\left\{ \min_{0 \leqslant y \leqslant \frac{1}{3}} (y^3 - 3y + 1), \quad \min_{\frac{1}{3} \leqslant y \leqslant 1} y^3 \right\}.$$

Evaluating the inner minima, we observe that each one of them is attained at $y = \frac{1}{3}$ and is equal to $\frac{1}{27}$. Moreover, $y^* = \frac{1}{3}$.

To determine the essential pure strategies for player I we form the equation

$$v = H(x, y^*), \quad \text{i.e. } \frac{1}{27} = \frac{1}{27} - x + x^3,$$

whence $x_1 = 0$ and $x_2 = 1$ (the third root of the above equation is $-1$ and has no game-theoretical meaning). Finally,

$$\left.\frac{\partial y^3}{\partial y}\right|_{y=\frac{1}{3}} = \frac{1}{3} > 0,$$

$$\left.\frac{\partial (y^3 - 3y + 1)}{\partial y}\right|_{y=\frac{1}{3}} = \frac{1}{3} - 3 = -\frac{8}{3} < 0.$$

Therefore $\alpha$ is determined from the equation

$$\alpha \cdot \tfrac{1}{3} + (1-\alpha)\left(-\tfrac{8}{3}\right) = 0, \quad \text{whence } \alpha = \tfrac{8}{9}.$$

We thus conclude that the optimal strategy for player I consists of choosing $x=0$ with probability $\frac{8}{9}$ and $x=1$ with probability $\frac{1}{9}$.

## 2.15 Market competition

*2.15.1.* Consider the game $\Gamma$ on the unit square with the following payoff function

$$H(x,y) = \begin{cases} k_1(x-y), & \text{if} \quad x \geqslant y \quad (k_1 > 0), \\ k_2(y-x), & \text{if} \quad x \leqslant y \quad (k_2 > 0). \end{cases} \tag{15.1}$$

This game can be interpreted as a competition between two firms for a market for their products under the capitalist system of economy.

Let one of the firms (player I) attempt to "squeeze" the second (player II)—who controls two markets for their product—out of *one* of the available markets. The total amount of funds allocated by player I for this purpose is assumed to be equal to one unit. Strategies for player I are the various allocations of funds between these two markets. If the amount $x$ is allocated to the first market, then $1-x$ is the sum of money assigned to the second. To keep his or her markets intact, player II also allocates one unit of funds and his or her strategies are to assign $y$ units to the first market and $1-y$ to the second.

We shall assume that player I, by achieving an advantage on one of the markets (clearly, he or she cannot conquer both markets simultaneously), eliminates his or her opponent from this market and obtains a payoff that is equal to the excess of his or her funds multiplied by a certain coefficient which reflects the importance of this market (we assume that the coefficient for the first market is $k_1$ and for the second is $k_2$).

It is clear that in formal notation the payoff function of the game just described is given by relationship (15.1).

*2.15.2.* We now show that the game is convex. For this purpose fix some $x=x_0$. The graph of $H(x_0,y)$ as a function in $y$ represents a pair of linear intervals as is shown in Figure 17 (if either $x_0=0$ or $x_0=1$ one of the

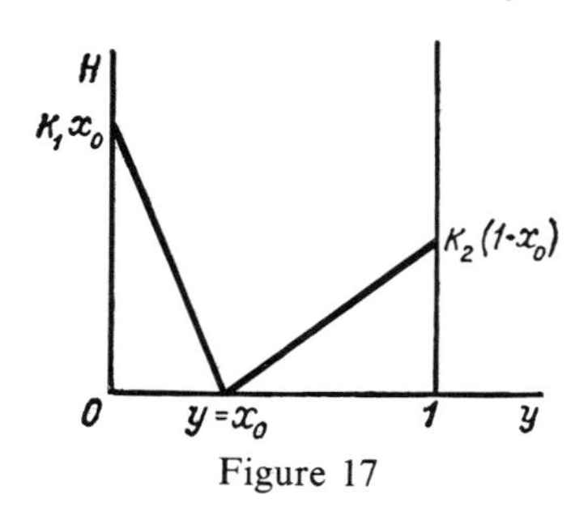


Figure 17

intervals shrinks to a point). Clearly, the function $H(x_0,y)$ is a convex function in $y$ for any $x_0$, so that the game with the payoff function $H(x,y)$ is indeed a convex game.

*2.15.3.* We have

$$\max_x H(x,y) = \max\Big\{\max_{x \geqslant y} k_1(x-y), \max_{x \leqslant y} k_2(x-y)\Big\}$$
$$= \max\{k_1(1-y), k_2 y\}.$$

Therefore

$$v = \min_y \max_x H(x,y) = \min_y \max_x \{k_1(1-y), k_2 y\}.$$

The graph of the function $\max\{k_1(1-y), k_2 y\}$ is indicated in Figure 18 by a boldface broken line. The first term to the right of the maximum sign is a decreasing function of $y$, while the second is an increasing one. Hence, for small values of $y$, the maximum is attained at the first term, while for large values of $y$ at the second. Hence, the minimal value of this maximum is attained for $y^*$ satisfying

$$k_1(1-y^*) = k_2 y^*,$$

i.e. for

$$y^* = \frac{k_1}{k_1+k_2}. \tag{15.2}$$

Thus, the obtained value of $y^*$ is the unique (pure) optimal strategy for player II. We see that the optimal strategy for player II consists of distributing his or her resources between the markets in the proportion corresponding to the importance of the markets.

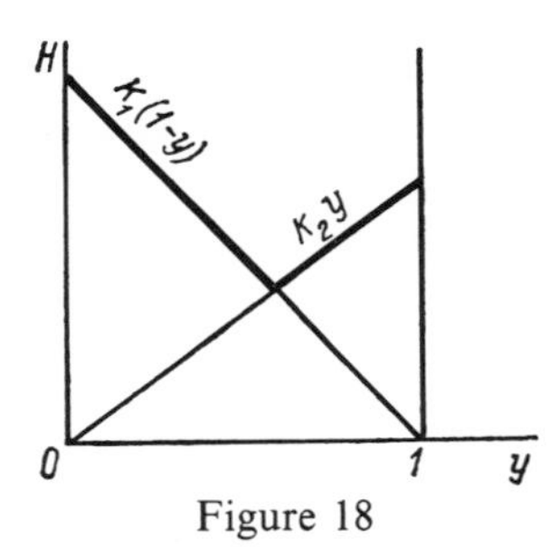


Figure 18

*2.15.4.* The value of the game is easily obtained to be equal to

$$v = \max H(x,y^*) = k_2 y^* = \frac{k_1 k_2}{k_1+k_2}.$$

*2.15.5.* To obtain an optimal strategy for player I, we first determine his or her essential strategies using equation (12.1). We shall consider separately the cases of $x \geqslant y^*$ and $x \leqslant y^*$.

For $x \geqslant y^*$, (12.1) becomes

$$k_1\left(x-\frac{k_1}{k_1+k_2}\right)=\frac{k_1k_2}{k_1+k_2};$$

hence $x=1$.

However, for $x \leqslant y^*$, (12.1) yields

$$k_2\left(\frac{k_1}{k_1+k_2}-x\right)=\frac{k_1k_2}{k_1+k_2};$$

thus $x=0$.

Whence, the essential strategies for player I are $x_1=0$ and $x_2=1$. In other words, player I possesses a unique optimal strategy which is a mixture of two pure strategies.

We now evaluate the required partial derivatives:

$$\left.\frac{\partial H(0,y)}{\partial y}\right|_{y=y^*}=\left.\frac{\partial}{\partial y}k_2y\right|_{y=y^*}=k_2>0,$$

$$\left.\frac{\partial H(1,y)}{\partial y}\right|_{y=y^*}=\left.\frac{\partial}{\partial y}k_1(1-y)\right|_{y=y^*}=-k_1<0.$$

Equation (12.6) for the given game reduces to

$$\alpha k_2+(1-\alpha)(-k_1)=0,$$

whence

$$\alpha=\frac{k_1}{k_1+k_2}.$$

In other words, the optimal strategy for player I is to concentrate all his or her resources on one of two markets and the probability of choosing this "market of concentration" is inversely proportional to its importance. The result is not surprising: the more important the market is, the more funds will presumably be allocated by the opponent in an effort to retain it; thus if the opponent is eliminated from this market the amount of *free* funds still remaining will be smaller and the victory for player I will be less significant.

## 2.16 Allocation of production capacities; minimization of the maximal intensity of a production scheme

*2.16.1.* Let the total production capacity of two enterprises of the same type scheduled for construction in two different locations be fixed (we shall assume that the total capacity is one unit). Also let the total production capacity be equal to the overall demand for the product produced by these enterprises at the two locations. The exact demand at each one of the locations is unknown (it can be determined only *after* the enterprises start their actual production). If the production capacity in one of the locations equals $y$ and the demand at this location is $x$, then the "intensity" of the performance by the enterprise is measured by the ratio $x/y$. The aim of an optimal distribution of the production capacities corresponds to the minimization of the maximal intensity of activities of both enterprises, i.e. minimization of the maximum

$$\max\left\{\frac{x}{y}, \frac{1-x}{1-y}\right\}. \tag{16.1}$$

Thus the problem posed can be described by a game in which player I (the "nature" or "circumstances") chooses a value $x$ and player II (the designer) chooses a value $y$ and the payoff function $H(x,y)$ (which describes the losses of the designer) is equal to (16.1).

Strictly speaking, the game is not defined on the whole unit square since for $y=0$ and $y=1$ the payoff function is not determined. However, if in our arguments we do not revert to these extreme pure strategies for player II, all the properties of convex games of the unit square discussed in previous sections remain valid for the game under consideration as well.

*2.16.2.* The game under consideration is convex. Indeed, fix some $x=x_0$. Then

$$H(x_0,y)=\max\left\{\frac{x_0}{y}, \frac{1-x_0}{1-y}\right\}.$$

The graph of this function is a pair of segments of two hyperbolas (see Figure 19). (As $x_0$ approaches 0 or 1, one of these segments shrinks to a point.) Therefore, $H(x_0,y)$ is a convex function and the game is a convex game (moreover, it is *strictly* convex).

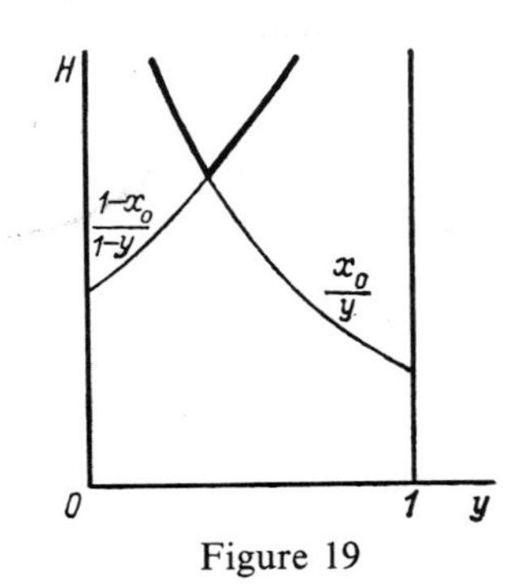


Figure 19

*2.16.3.* To determine the optimal strategy for player II (which in our case is unique: cf. Section 2.12), we shall consider the minimum of the expression

$$\max_{0 \leqslant x \leqslant 1} H(x,y) = \max_{0 \leqslant x \leqslant 1} \max\left\{\frac{x}{y}, \frac{1-x}{1-y}\right\},$$

or utilizing the fact that the order of two extremum operations of the *same* kind can be interchanged (this is not valid for extremum operations of the "opposing" kind—maximization and minimization), we obtain

$$\max_{0 \leqslant x \leqslant 1} H(x,y) = \max\left\{ \max_{0 \leqslant x \leqslant 1} \frac{x}{y}, \max_{0 \leqslant x \leqslant 1} \frac{1-x}{1-y} \right\}$$
$$= \max\left\{ \frac{1}{y}, \frac{1}{1-y} \right\}.$$

The minimum of the last expression is attained at the optimal strategy $y^*$ for player II satisfying

$$\frac{1}{y^*} = \frac{1}{1-y^*},$$

i.e. $y^* = \frac{1}{2}$.

Consequently, the optimal strategy for player II will be to subdivide the production capacities into two equal parts.

*2.16.4.* The value of the game is therefore equal to

$$v = \max_x H(x,y^*) = \tfrac{1}{2}.$$

*2.16.5.* The essential strategies for player I are $x=0$ and $x=1$. To determine the optimal strategy for this player, we evaluate the partial derivatives

$$\left.\frac{\partial H(0,y)}{\partial y}\right|_{y=y^*} = \frac{1}{(1-y^*)^2} = \frac{1}{4} > 0,$$
$$\left.\frac{\partial H(1,y)}{\partial y}\right|_{y=y^*} = -\frac{1}{y^{*2}} = -\frac{1}{4} < 0.$$

We have

$$\tfrac{1}{4}\alpha - \tfrac{1}{4}(1-\alpha) = 0, \quad \text{whence } \alpha = \tfrac{1}{2}.$$

We thus observe that the least favorable "situation" for player II (the designer) is when all the demand is concentrated in one location, with equal probability for either of them. In this case, one of the enterprises will be idle, while the other will be working with double intensity.

## 2.17 Allocation of production capacities under partial uncertainty

*2.17.1.* If the designer in the problem discussed in the previous section had known the actual distribution of the total demand (i.e. the value of $y$), he or she would have chosen $x = y$ and each of the enterprises would have been working at full capacity without an overload. However, as it was shown above in the case when no definite information is available about the demand, one cannot escape the possibility of a double load. It is natural to assume that if the designer possesses partial information concerning the distribution of the demand, he may have a smaller overload and the narrower the region in which the possible distribution of demand is situated, the smaller the amount of overload will be.

Let it be known to the designer that the demand in the first location can vary within the segment $[a,b]$, where $0 \leqslant a \leqslant b \leqslant 1$. In this case, we are dealing with a game in which the set of strategies for player I is the segment $[a,b]$ and the set of strategies for player II is the *open* interval $(0,1)$. The form of the payoff function is the same as in the game discussed in the preceding section:

$$H(x,y) = \max\left\{\frac{x}{y}, \frac{1-x}{1-y}\right\}.$$

As it was already pointed out, the distinction between our present game and the game on the unit square is inconsequential and we can consider our game as a game on the unit square. Clearly the game is convex, even strictly convex.

*2.17.2.* The unique optimal strategy for player II is the pure strategy at which the minimum

$$\begin{aligned}\min_{0<y<1} \max_{a \leqslant x \leqslant b} H(x,y) &= \min_{0<y<1} \max_{a \leqslant x \leqslant b} \max\left\{\frac{x}{y}, \frac{1-x}{1-y}\right\} \\ &= \min_{0<y<1} \max\left\{\max_{a \leqslant x \leqslant b} \frac{x}{y}, \max_{a \leqslant x \leqslant b} \frac{1-x}{1-y}\right\} \\ &= \min_{0<y<1} \max\left\{\frac{b}{y}, \frac{1-a}{1-y}\right\}\end{aligned}$$

is attained.

Clearly, this minimum is attained at $y^*$ satisfying the equation

$$\frac{b}{y^*} = \frac{1-a}{1-y^*},$$

i.e. for

$$y^* = \frac{b}{1+b-a}. \tag{17.1}$$

*2.17.3.* In the present case

$$v = 1 + (b-a), \tag{17.2}$$

i.e. the expected *additional* overload is equal to $b-a$ (in the particular case discussed in the preceding section $b-a=1$, so that the extra overload was equal to the initial "load").

*2.17.4.* To determine the essential strategies for player I we observe that it is easy to verify that in view of (17.1), $a \leqslant y^* \leqslant b$, with the equality attained only if $a = b$. Since the latter case corresponds to a complete determinicity of the demand, which is of no interest from the game-theoretical approach, we shall assume that $a < y^* < b$.

Clearly, the essential strategies for player I are $x = a$ and $x = b$. To determine the probabilities with which they occur in the unique optimal mixed strategy we evaluate the derivatives

$$\left.\frac{\partial H(a,y)}{\partial y}\right|_{y=y^*} = \frac{1-a}{(1-y^*)^2} = \frac{(1+b-a)^2}{1-a} > 0,$$

$$\left.\frac{\partial H(b,y)}{\partial y}\right|_{y=y^*} = -\frac{b}{y^{*2}} = -\frac{(1+b-a)^2}{b} < 0,$$

and the required probabilities are determined from the equation

$$\alpha\frac{(1+b-a)^2}{1-a} - (1-\alpha)\frac{(1+b-a)^2}{b} = 0,$$

whence

$$\alpha = \frac{1-a}{1+b-a}. \tag{17.3}$$

*2.17.5.* Assume, for example, that the designer knows that the demand in the first location may vary from 30 to 60% of the total amount, i.e. that $a = 0.3$ and $b = 0.6$. According to (17.1) he should allocate at that location $0.6/(1+0.6-0.3) = 0.46$, i.e. 46% of the production capacity. The overload coefficient determined by the value of the game according to (17.2) will then be $1+0.6-0.3 = 1.3$ and the least favorable "situation," in view of (17.3), is when the demand at the first location is the total overall demand (in the amount of $1+0.6 = 1.6$) with probability

$$\frac{1-0.3}{1+0.6-0.3} = 0.54.$$

# 3 Noncooperative games

## 3.1 Mixed extensions of noncooperative games

*3.1.1.* In the first few sections of Chapter I we introduced the definitions of a noncooperative game and an equilibrium situation. The rationale behind the desire and tendency of the players to attain equilibrium situations was explained. Such a tendency towards equilibrium can thus be viewed as a kind of optimal behavior. A very important fact associated with these concepts is that an equilibrium situation is indeed attainable for a substantially wide class of noncooperative games. Hence the notion of an equilibrium situation is not only logically sound but is also practically useful.

Our study of antagonistic games has revealed that a large number of these games possess an equilibrium situation in mixed rather than in pure strategies. Hence, in the case of more general noncooperative games, it is also reasonable to search for equilibrium situations in mixed strategies.

*3.1.2.* For this purpose we introduce the notion of a mixed extension of a noncooperative game. Let

$$\Gamma = \langle I, \{S_i\}_{i\in I}, \{H_i\}_{i\in I}\rangle$$

be an arbitrary noncooperative game. We assume that the game is finite, i.e. that the set $S_i$ of pure strategies for each one of the players is a finite set.

Let $\sigma_i$ be an arbitrary mixed strategy for the $i$th player, i.e. a certain probability distribution on the set $\{S_i\}$. The probability assigned by $\sigma_i$ to the pure strategy $s_i$ is denoted by $\sigma_i(s_i)$. The set of all the mixed strategies of player $i$ is denoted by $\Sigma_i$.

Let each of the players $i \in I$ apply his or her mixed strategy $\sigma_i$, i.e chooses its pure strategies $s_i$ with probabilities $\sigma(s_i)$. Moreover, let the mixed strategies of all the players $1,\ldots,n$—viewed as probability distributions—be *jointly independent*, i.e. the probability of arriving at the situation $s=(s_1,\ldots,s_n)$ is assumed to be the *product* of probabilities of choosing its components: $\sigma_1(s_1)\sigma_2(s_2)\cdots\sigma_n(s_n)$.

Thus we arrive at the probability distribution $\sigma$ on the set of all situations defined by

$$\sigma(s)=\sigma(s_1,\ldots,s_n)=\sigma_1(s_1)\cdots\sigma_n(s_n)$$

for all situations in the game $\Gamma$. These types of probability distributions $\sigma$ are called *situations* of game $\Gamma$ in *mixed strategies*.

A situation of the game $\Gamma$ in mixed strategies is the realization of various actual situations in pure strategies, each occurring with a certain probability. Therefore the value of the payoff function for each of the players will be a random variable. In game theory the value of the payoff function for player $i$ in a situation in mixed strategies is stipulated to be the mathematical expectation of this random variable, i.e.,

$$H_i(\sigma)=\sum_{s\in S} H_i(s)\sigma(s)=\sum_{s_1\in S_1}\cdots\sum_{s_n\in S_n} H_i(s_1,\ldots,s_n)\prod_{i=1}^{n}\sigma_i(s_i). \tag{1.1}$$

We note, interalia, that

$$H_i(\sigma\|s_j^0)=\sum_{s_1\in S_1}\cdots\sum_{s_{j-1}\in S_{j-1}}\sum_{s_{j+1}\in S_{j+1}}\cdots\sum_{s_n\in S_n} H_i(s\|s_j^0)\prod_{\substack{i=1\\ i\neq j}}^{n}\sigma_i(s_i). \tag{1.2}$$

(The notation $\|$ was defined in Section 1.2.1.)

*3.1.3* **Definition.** The game

$$\Gamma^*=\langle I,\{\Sigma_i\}_{i\in I},\{H_i\}_{i\in I}\rangle,$$

in which the set of players is $I$, the set of strategies for the player $i$ is $\Sigma_i$ (for each $i\in I$), and the payoff function is defined by equality (1.1), is called a *mixed extension of the game* $\Gamma$.

*3.1.4.* In general noncooperative games it is possible to pass on to mixed strategies in a similar manner as in antagonistic games. This was described in Sections 1.10.3 and 2.4.5. Namely, if for any pure strategy $s_i$ for player $i$ and some number $a$ the inequality $H_j(\sigma\|s_i)\leqslant a$ is valid, then for any mixed strategy $\sigma_i^*$ of this player the inequality $H_j(\sigma\|\sigma_i^*)\leqslant a$ is also satisfied.

The proof of this assertion follows from equations (1.1) and (1.2) using the standard transition to mixed strategies.

*3.1.5.* Later the assertion stated below will be required.

**Lemma.** *For any situation in mixed strategies $\sigma=(\sigma_1,\ldots,\sigma_n)$ each of the players $i\in I$ possesses a pure strategy $s_i^0$ such that the following two* *inequalities are simultaneously satisfied*:

$$\sigma_i(s_i^0) > 0, \tag{1.3}$$

*and*

$$H_i(\sigma \| s_i^0) \leqslant H_i(\sigma).$$

PROOF. Suppose that for all pure strategies $s_i$ for player $i$ satisfying (1.3) we have

$$H_i(\sigma \| s_i) > H_i(\sigma).$$

Then for all these strategies

$$H_i(\sigma \| s_i)\sigma_i(s_i) > H_i(\sigma)\sigma_i(s_i). \tag{1.4}$$

This inequality holds for all $s_i$ such that $\sigma_i(s_i) > 0$. However, for all other $s_i$ we have

$$H_i(\sigma \| s_i)\sigma_i(s_i) = H_i(\sigma)\sigma_i(s_i) = 0. \tag{1.5}$$

Adding up all the inequalities (1.4) and all the equations (1.5) we sum up over *all* pure strategies $s_i$ for player $i$ and obtain

$$\sum_{s_i \in S_i} H_i(\sigma \| s_i)\sigma_i(s_i) > \sum_{s_i \in S_i} H_i(\sigma)\sigma_i(s_i),$$

or utilizing (1.1) and (1.2) we obtain

$$H_i(\sigma) > H_i(\sigma),$$

which is a contradiction.

This contradiction establishes the existence of the required pure strategy for player $i$. □

## 3.2 Equilibrium situations

*3.2.1.* According to the definition presented in Section 1.2, an equilibrium situation in a noncooperative game

$$\Gamma = \langle I, \{S_i\}_{i \in I}, \{H_i\}_{i \in I} \rangle$$

is a situation $s^*$ such that for each $i \in I$ and $s_i \in S$

$$H_i(s^* \| s_i) \leqslant H_i(s^*).$$

**Definition.** An equilibrium situation in a mixed extension $\Gamma^*$ of the game $\Gamma$ is called *an equilibrium situation of the game* $\Gamma$ *in mixed strategies*.

Thus a situation $\sigma^*$ is an equilibrium situation in the game $\Gamma^*$ if for any player $i$ and for any mixed strategy $\sigma_i$ for this player the inequality

$$H_i(\sigma^* \| \sigma_i) \leqslant H_i(\sigma^*) \tag{2.1}$$

is valid.

As it was done in the case of antagonistic games, mixed strategies for players are often referred to simply as strategies, specifying that a strategy is pure when necessary. We shall adhere to this terminology also in relation to *situations* in mixed strategies and in particular to equilibrium situations in mixed strategies.

*3.2.2.* The following theorem is often very useful.

**Theorem.** *In order that a situation* $\sigma^*$ *in the game* $\Gamma$ *be an equilibrium situation for this game (in mixed strategies) it is necessary and sufficient that for any player* $i$ *and any pure strategy* $s_i$ *for this player the inequality*

$$H_i(\sigma^* \| s_i) \leqslant H_i(\sigma^*) \tag{2.2}$$

*be satisfied.*

PROOF. Necessity is obvious, since a pure strategy is a particular case of a mixed one and hence each one of the inequalities (2.2) (for each $i$) is a particular case of the corresponding inequality in (2.1).

To prove sufficiency, choose an arbitrary mixed strategy $\sigma_i$ for player $i$, multiply the inequality (2.2) by $\sigma_i(s_i)$, and sum up with respect to $s_i \in S_i$. This results in

$$\sum_{s_i \in S_i} H_i(\sigma^* \| s_i)\sigma_i(s_i) \leqslant \sum_{s_i \in S_i} \sigma_i(s_i)H_i(\sigma^*).$$

Applying (1.1) and (1.2) on the left-hand side of the last inequality and taking outside the summation sign on the right-hand side the expression $H_i(\sigma^*)$, which is independent of $s_i$, we obtain the required inequality (2.2). □

## 3.3 Nash's theorem

*3.3.1.* J. F. Nash* proved the existence of equilibrium situations in mixed strategies for any finite noncooperative game.

**Theorem.** *In any noncooperative game* $\Gamma = \langle I, \{S_i\}_{i \in I}, \{H_i\}_{i \in I} \rangle$ *there is at least one equilibrium situation (in mixed strategies).*

PROOF. If player $i$ in the game $\Gamma$ possesses $m_i$ pure strategies, then the set $\Sigma_i$ of all his or her mixed strategies (as was mentioned before) represents geometrically an $(m_i - 1)$-dimensional simplex. We shall denote this simplex by $S^{(i)}$. Therefore any situation

$$\sigma = (\sigma_1, \ldots, \sigma_n) \tag{3.1}$$

in mixed strategies can be viewed as a point in the Cartesian product $S^{(1)} \times \cdots \times S^{(n)}$ of these simplices of mixed strategies. This Cartesian product is clearly a convex closed bounded subset of the Euclidean space of dimensionality $m_1 + \cdots m_n - n$.

* *Ann. of Math*, **54** (1951), 286–295.

Now we define for an arbitrary situation $\sigma$ and any pure strategy $s_i^{(j)} \in S_i$ for player $i$

$$\phi_{ij}(\sigma) = \max\{0, H_i(\sigma \| s_i^{(j)}) - H_i(\sigma)\}. \tag{3.2}$$

The functions $\phi_{ij}$ defined by (3.2) take on only nonnegative values.

These functions measure the increment in the payoff for player $i$ in a situation $\sigma$ due to the replacement of his strategy $\sigma_i$—which appears in this situation—by a certain pure strategy $s_i^{(j)}$. The possible decrease due such a replacement is not indicated by function $\phi_{ij}$ since in the latter case the function $\phi_{ij}$ vanishes.

Now construct for all possible $i = 1, \ldots, n$ and $j = 1, \ldots, m_i$ numbers of the form

$$\frac{\sigma_i(s_i^{(j)}) + \phi_{ij}(\sigma)}{1 + \sum_{j=1}^{m_i} \phi_{ij}(\sigma)}. \tag{3.3}$$

Clearly, all these fractions are nonnegative and each sum of the form

$$\sum_{j=1}^{m_i} \frac{\sigma_i(s_i^{(j)}) + \phi_{ij}(\sigma)}{1 + \sum_{j=1}^{m_i} \phi_{ij}(\sigma)}$$

is equal to 1.

Consequently, for fixed $\sigma$ and $i$ the fractions (3.3) can be viewed as probabilities of the corresponding pure strategies $s_i^{(j)}$ for player $i$. A collection of these fractions for all pure strategies $s_i^{(j)}$ can thus be interpreted as a mixed strategy for player $i$.

Since the fractions (3.3) are constructed for each $i$, their totality determines a system of mixed strategies for all the players, i.e. a situation in game $\Gamma$. Such a situation is a function of the initial situation $\sigma$. It will be denoted by $f(\sigma)$. Clearly the function $f$ maps the closed convex and bounded set of all situations $\Sigma$ into itself.

Moreover, this function is a continuous function of situations. Indeed, each component of the situation that is the value of function $f$ is a fraction of the form (3.3). The first summand in the numerator of this fraction is the component of the initial situation and hence the dependence here is continuous. According to (3.2) the second summand consists of linear functions $H_i(\sigma)$ and $H_i(\sigma \| s_i^{(j)})$, the constant 0, and the maximization operator. (Observe that the function $\max\{0, x\}$ is a continuous function $x$ as it is shown in the graph presented in Figure 20.) Consequently, $\phi_{ij}(\sigma)$ are also continuous functions of $\sigma$. Hence the numerator of the fraction (3.3) is a continuous function of $\sigma$. Finally, the denominator of this fraction is continuous and does not vanish (since its value is at least 1). Thus the function $f$ is indeed a continuous function.

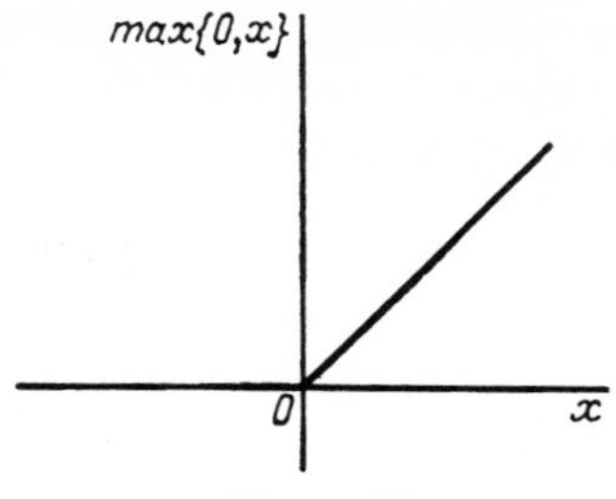


Figure 20

In view of the above, the conditions of the well-known *Brouwer's fixed point theorem** are satisfied. This theorem asserts that a continuous mapping $f$ of a convex subset of a finite-dimensional space into itself possesses at least one fixed point, i.e. a point $\sigma^0$ such that $f(\sigma^0)=\sigma^0$. Let $\sigma^0$ be one of the fixed points. This implies that for all $i$ and $j$

$$\sigma_i^0(s_i^{(j)}) = \frac{\sigma_i^0(s_i^{(j)}) + \phi_{ij}(\sigma^0)}{1 + \sum_{j=1}^{m_i} \phi_{ij}(\sigma^0)}. \tag{3.4}$$

Recall now that in view of the lemma in Section 3.1.5 there exists for any player $i$ a pure strategy $s_i^0$ such that $\sigma_i^0(s_i^0)>0$ and $\phi_{i0}(\sigma^0)=0$. For this particular strategy the equality (3.4) becomes

$$\sigma_i^0(s_i^0) = \frac{\sigma_i^0(s_i^0) + \phi_{i0}(\sigma^0)}{1 + \sum_{j=1}^{m_i} \phi_{ij}(\sigma^0)},$$

whence

$$\sigma_i^0(s_i^0) + \sigma_i^0(s_i^0) \sum_{j=1}^{m_i} \phi_{ij}(\sigma^0) = \sigma_i^0(s_i^0) + \phi_{i0}(\sigma^0).$$

Subtracting the equal summands from both sides of the last equality and noting that the second summand on the right-hand side vanishes (in view of the choice of $s_i^0$), we obtain

$$\sigma_i^0(s_i^0) \sum_{j=1}^{m_i} \phi_{ij}(\sigma^0) = 0.$$

However, for the very same reason, the first factor on the right-hand side is *nonzero*. Hence

$$\sum_{j=1}^{m_i} \phi_{ij}(\sigma^0) = 0.$$

*There are numerous proofs of Brouwer's fixed point theorem. One of the most elementary is given in K. Kuga, Brouwer's fixed point theorem: an alternative proof, *Siam Journ. of Appl. Math.*, **5** (1974), 893–897 (Translator's remark).

Furthermore, since all the numbers $\phi_{ij}(\sigma^0)$ are nonnegative, it follows from the last equation that $\phi_{ij}(\sigma^0)=0$ for each of them. Thus, in this case there are no positive numbers in equality (3.2) to the right of the max sign; in other words

$$H_i(\sigma^0 \| s_i^{(j)}) \leqslant H_i(\sigma^0).$$

Since this inequality is satisfied for each player $i$ and any pure strategy $s_i^{(j)}$ for this player, the theorem in Section 3.2 implies that the situation $\sigma^0$ is an equilibrium situation.

*3.3.2.* Observe that the basic importance of Nash's theorem is in assuring the existence of an equilibrium situation. The theorem, however, cannot be applied for actual determination of such a situation; this is not a "constructive" result because Brouwer's fixed point theorem does not reveal how to find a fixed point but only guarantees its existence. Consequently, Nash's theorem, which is totally based on Brouwer's, does not show how one finds an equilibrium situation.

## 3.4 Properties of equilibrium situations

*3.4.1.* As it was observed in Section 1.7.2, the set of equilibrium situations in antagonistic games (i.e. the set of saddle points) forms a rectangular set. Moreover, the players in an antagonistic game have the same payoff at all the equilibrium situations (the payoffs for the two opponents are of the same value but of opposing sign). The following example shows that in general noncooperative games neither of these properties may be valid.

EXAMPLE. Consider a noncooperative game

$$\Gamma=\langle I,\{S_i\}_{i\in I},\{H_i\}_{i\in I}\rangle,$$

where

$$I=\{1,2,\ldots,n\}, \qquad S_i=\{1,2,\ldots,m\} \quad \text{for all } i\in I,$$

and

$$H_i(s)=\begin{cases} a_{ik}>0, & \text{if } s_1=s_2=\ldots=s_n=k,\\ 0 & \text{otherwise.}\end{cases}$$

If we set $s^k=(k,k,\ldots,k)$, then for each $i\in I$,

$$H_i(s^k\|s_i)=0 \quad \text{for any } s_i\neq k.$$

Therefore

$$H_i(s^k\|s_i)\leqslant H_i(s^k),$$

i.e. $s^k$ is an equilibrium situation. Moreover it is easy to verify that there are no other equilibrium situations in *pure strategies* in this game. Hence the set of equilibrium situations in the game $\Gamma$ is not a rectangular set.

On the other hand, if the numbers $a_{ik}$ for the same $i$ but different $k$ are distinct, then player $i$ will have different payoffs in different equilibrium situations.

In view of the above, for general noncooperative games there exist no simple and convenient notions such as an "optimal" strategy for a player or the "value" of a game. Moreover, a collection of equilibrium strategies for different players do not always form an equilibrium situation (only certain combinations of equilibrium strategies do). This fact diminishes the importance of an equilibrium situation in general noncooperative games and also impedes their determination.

*3.4.2.* However, certain properties of saddle points in antagonistic games are also valid for equilibrium situations in arbitrary noncooperative games. The following simple but important assertion is one of them.

**Theorem.** *If an equilibrium strategy $\sigma_i$ for player i appears in an equilibrium situation $\sigma$ and if for some pure strategy $s_i^0$ for this player the strict inequality*

$$H_i(\sigma \| s_i^0) < H_i(\sigma)$$

*is satisfied, then $\sigma_i(s_i^0) = 0$.*

The proof of this theorem is basically the same as that of the analogous result in Section 1.20.8:

Assume that $\sigma_i(s_i^0) > 0$. Then

$$H_i(\sigma \| s_i^0)\sigma_i(s_i^0) < H_i(\sigma)\sigma_i(s_i^0). \tag{4.1}$$

However, for all pure strategies for player $i$ different from $s_i^0$ we have

$$H_i(\sigma \| s_i) \leqslant H_i(\sigma)$$

according to the definition of an equilibrium situation; whence

$$H_i(\sigma \| s_i)\sigma_i(s_i) \leqslant H_i(\sigma)\sigma_i(s_i). \tag{4.2}$$

Summing up over all $s_i \neq s_i^0$, adding (4.1) to this sum, and utilizing the argument used in the proof of Nash's theorem, we see that $H_i(\sigma) < H_i(\sigma)$, which is a contradiction. □

*3.4.3.* The theorem just proved can be stated in a somewhat different form: *if an equilibrium strategy $\sigma_i$ for player i appears in the equilibrium situation $\sigma$ and the pure strategy $s_i^0$ is essential in the mixed strategy $\sigma_i$, then*

$$H_i(\sigma \| s_i^0) = H_i(\sigma). \tag{4.3}$$

## 3.5 Bi-matrix games

*3.5.1.* At present there are still no general methods available to determine equilibrium situations applicable to arbitrary finite noncooperative games. It seems that mathematically this problem is extremely difficult. However, for some individual sufficiently simple classes of noncooperative games this problem can be solved.

One of such classes is the class of finite noncooperative two-person games. Let player I in such a game possess $m$ pure strategies and player II $n$ pure strategies and let the payoff to player I in situation $(i,j)$ $(i=1,\ldots,m;$ $j=1,\ldots,n)$ be $a_{ij}$ while the corresponding payoff to player II be $b_{ij}$. The payoff functions for both players are then conveniently expressed by a pair of matrices

$$\mathbf{A}=\begin{bmatrix} a_{11}\ldots a_{1n} \\ \vdots \\ a_{m1}\ldots a_{mn} \end{bmatrix}, \qquad \mathbf{B}=\begin{bmatrix} b_{11}\ldots b_{1n} \\ \vdots \\ b_{m1}\ldots b_{mn} \end{bmatrix}.$$

This is the reason that the name *bi-matrix* is used for this type of game.

*3.5.2.* Mixed strategies in bi-matrix games (analogously to the case of matrix games) can be interpreted as vectors forming a fundamental simplex. If $X$ and $Y$ are vectors representing mixed strategies for players I and II, then it is easy to verify that

$$H_{\mathrm{I}}(X,Y)=X\mathbf{A}Y^T, \qquad H_{\mathrm{II}}(X,Y)=X\mathbf{B}Y^T.$$

An equilibrium situation in a bi-matrix game can be defined in the following manner. Situation $(X,Y)$ in a bi-matrix game with the payoff matrices $\mathbf{A}$ and $\mathbf{B}$ is an equilibrium situation if

$$A_{i\cdot}Y^T \leqslant X\mathbf{A}Y^T, \qquad i=1,\ldots,m, \tag{5.1}$$

$$XB_{\cdot j} \leqslant X\mathbf{B}Y^T, \qquad j=1,\ldots,n. \tag{5.2}$$

Clearly, if $\mathbf{B}=-\mathbf{A}$, a bi-matrix game reduces to a matrix game and in that case relations (5.1) and (5.2) will be correspondingly

$$A_{i\cdot}Y^T \leqslant X\mathbf{A}Y^T, \qquad i=1,\ldots,m, \tag{5.3}$$

$$-XA_{\cdot j} \leqslant -X\mathbf{A}Y^T, \qquad j=1,\ldots,n. \tag{5.4}$$

The last inequality (5.4) is equivalent to

$$X\mathbf{A}Y^T \leqslant XA_{\cdot j},$$

which together with (5.3) gives us the definition of a saddle point.

## 3.6 Solutions of bi-matrix games

*3.6.1.* A complete enumeration of all equilibrium situations for arbitrary bi-matrix games is possible although it is a very cumbersome task.

We shall therefore limit ourselves to a description of a method to determine equilibrium situations for the simplest bi-matrix games, namely for the $2\times 2$ games in which each player possesses only two pure strategies.

*3.6.2.* In what follows we shall consider a bi-matrix game with the payoff matrices

$$\mathbf{A}=\begin{pmatrix} a_{11} & a_{12} \\ a_{21} & a_{22} \end{pmatrix}, \qquad \mathbf{B}=\begin{pmatrix} b_{11} & b_{12} \\ b_{21} & b_{22} \end{pmatrix}$$

for players I and II, respectively. Mixed strategies $X$ and $Y$ for the players are uniquely determined by the probabilities $x$ and $y$ with which these players choose their first pure strategy (the second pure strategy is then automatically chosen with probabilities $1-x$ and $1-y$, respectively).

Since $x$ and $y$ are both in the closed interval between 0 and 1 (including the end points), any situation in a bi-matrix $2\times 2$ game is uniquely determined by a point $(x,y)$ of the unit square. The payoffs for the players in such a situation will be denoted for convenience by $H_1(x,y)$ and $H_2(x,y)$, respectively.

Clearly,

$$\begin{aligned} H_1(x,y) = X\mathbf{A}Y^T &= (a_{11}-a_{12}-a_{21}+a_{22})xy \\ &\quad +(a_{12}-a_{22})x+(a_{21}-a_{22})y+a_{22}, \end{aligned} \tag{6.1}$$

$$\begin{aligned} H_2(x,y) = X\mathbf{B}Y^T &= (b_{11}-b_{12}-b_{21}+b_{22})xy \\ &\quad +(b_{12}-b_{22})x+(b_{21}-b_{22})y+b_{22}. \end{aligned} \tag{6.2}$$

*3.6.3.* An equilibrium situation in a noncooperative game represents a situation that is admissible for each of the players. We shall therefore describe—in the unit square of all situations—the "geometric loci" of the situations which are admissible for each of the players separately.

We start with situations admissible for player I. In order that a situation $(x,y)$ be admissible for this player, it is necessary and sufficient that

$$H_1(1,y) = A_{1.}Y^T \leqslant X\mathbf{A}Y^T = H_1(x,y),$$
$$H_1(0,y) = A_{2.}Y^T \leqslant X\mathbf{A}Y^T = H_1(x,y)$$

be satisfied, or utilizing (6.1) and writing the payoffs explicitly,

$$\begin{aligned} &(a_{11}-a_{12}-a_{21}+a_{22})y+(a_{12}-a_{22})+(a_{21}-a_{22})y+a_{22} \\ &\qquad \leqslant (a_{11}-a_{12}-a_{21}+a_{22})xy+(a_{12}-a_{22})x+(a_{21}-a_{22})y+a_{22}, \\ &(a_{21}-a_{22})y+a_{22} \leqslant (a_{11}-a_{12}-a_{21}+a_{22})xy+(a_{12}-a_{22})x \\ &\qquad\qquad +(a_{21}-a_{22})y+a_{22}. \end{aligned}$$

After some elementary algebraic simplifications these inequalities become

$$(a_{11}-a_{12}-a_{21}+a_{22})(1-x)y+(a_{12}-a_{22})(1-x)\leqslant 0, \tag{6.3}$$

$$(a_{11}-a_{12}-a_{21}+a_{22})xy+(a_{12}-a_{22})x\geqslant 0. \tag{6.4}$$

To simplify the notation we denote

$$a_{11}-a_{12}-a_{21}+a_{22}=A, \qquad a_{22}-a_{12}=a.$$

(As we shall see below the quantities $A$ and $a$ are invariants for the set of admissible situations for player I in the class of $2\times 2$ bi-matrix games in the sense that this set is completely determined by the values of $A$ and $a$.)

With this notation inequalities (6.3) and (6.4) become

$$A(1-x)y-a(1-x)\leqslant 0, \tag{6.5}$$

$$Axy-ax\geqslant 0. \tag{6.6}$$

In other words, the set of all the admissible situations for player I is the intersection of the set of solutions of the system (6.5), (6.6) with the unit square $[0,1]\times[0,1]$.

*3.6.4.* We shall first describe all the solutions of the system (6.5), (6.6) located in the strip $[0,1]\times(-\infty,+\infty)$ and enumerate separately solutions of this system with $x=0$, $x=1$, and $0<x<1$.

If $x=0$, inequality (6.6) is automatically satisfied and inequality (6.5) becomes

$$Ay-a\leqslant 0. \tag{6.7}$$

If $x=1$, we have a symmetric situation: (6.5) becomes an identity and (6.6) reduces to

$$Ay-a\geqslant 0. \tag{6.8}$$

Now let $0<x<1$. The inequalities (6.5) and (6.6) can in this case be divided by $(1-x)$ and $x$, respectively, which yields the equalities (6.7) and (6.8), i.e. the equality

$$Ay-a=0. \tag{6.9}$$

Thus the set of all solutions of the system (6.5), (6.6) in the strip $[0,1]\times(-\infty,+\infty)$ consists of:

(a) all the situations of the form $(0,y)$, where $Ay-a\leqslant 0$;

(b) all the situations of the form $(x,y)$, where $x\in[0,1]$ and $Ay-a=0$;

(c) all the situations of the form $(1,y)$, where $Ay-a>0$.

*3.6.5.* The structure of this set depends on the values of the invariants $A$ and $a$.

If $A=a=0$, all the solutions are of type (b) and their totality encompasses the whole strip $[0,1]\times(-\infty,+\infty)$. Clearly, the set of all admissible situations for player I is in this case the whole unit square of situations.

If $A=0$, but $a\neq 0$, the solutions in the strip are either of type (a) or (c) depending on the sign of $a$. In this case the set of solutions of the system

(6.5), (6.6) is either the line $x=0$ or the line $x=1$ and, correspondingly, the set of all admissible situations is either the left-hand side of the unit square of the situation or the right-hand side.

Finally let $A\neq 0$. In this case all the solutions of the system (6.5), (6.6) of the form $(0,y)$ satisfy

$$y\leqslant a/A=\alpha, \quad \text{if } A>0,$$
$$y\geqslant a/A=\alpha, \quad \text{if } A<0.$$

This means that the set of values of $y$ is either the half-line $(-\infty,\alpha]$ or the half-line $[\alpha,\infty)$.

Solutions of (6.5), (6.6) of the form $(1,y)$ are symmetric to those of the form $(0,y)$. These solutions satisfy

$$y\geqslant \alpha, \quad \text{if } A>0,$$
$$y\leqslant \alpha, \quad \text{if } A<0,$$

i.e. the set of $y$ in this case is again either the half-line $[\alpha,\infty)$ or $(-\infty,\alpha]$.

Finally, for solutions of (6.5), (6.6) of the form $(x,y)$ with $0<x<1$ we have

$$y=a/A=\alpha, \tag{6.10}$$

i.e. the set of these solutions represents the segment joining the points $(0,\alpha)$ and $(1,\alpha)$.

Hence the set of *all* the solutions of (6.5), (6.6) for $A\neq 0$ is a zigzag. For the case $A>0$ this zigzag is represented in Figure 21 and for $A<0$ in Figure 22. Recall that the set of all admissible situations for player I is the intersection of this zigzag with the unit square. One verifies directly that for $\alpha<0$ the set of admissible situations is one of the *vertical* sides of the unit square; for $\alpha=0$ the set consists of two sides that form a right angle; for $0<\alpha<1$ it is a zigzag; for $\alpha=1$ it is again a right angle; and for $\alpha>1$ it is the other vertical side of the unit square of the situations.

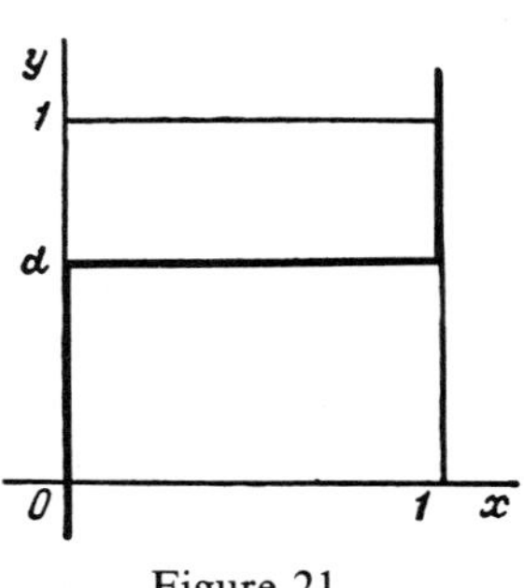


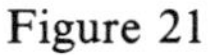
Figure 21

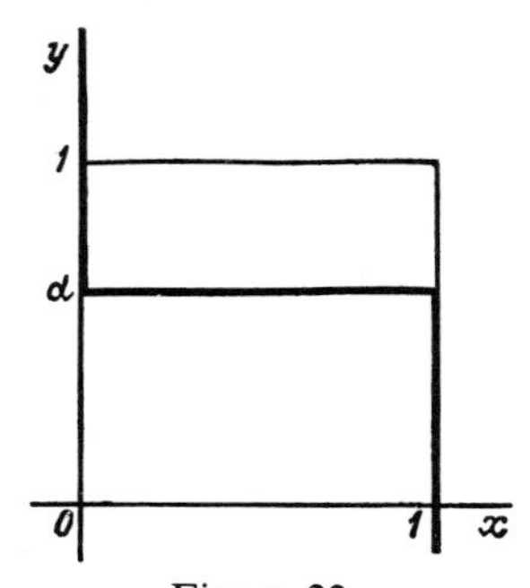


Figure 22

*3.6.6.* Enumeration of all the admissible situations for player II is carried out similarly. First, the "invariants" $B$ and $b$ for this set are calculated:

$$b_{11}-b_{12}-b_{21}+b_{22}=B, \qquad b_{22}-b_{21}=b. \tag{6.11}$$

The set of all the admissible situations for player II consists of:

(a) all the situations of the form $(x,0)$, where $Bx-b<0$;

(b) all the situations of the form $(x,y)$, where $Bx-b=0$ and $y\in[0,1]$;

(c) all the situations of the form $(x,1)$, where $Bx-b>0$.

If $B=b=0$, then every situation in the game will be admissible for player II.

However, if $B=0$, but $b\neq 0$ the set of all the admissible situations for player II will either be the lower or the upper side of the unit square of situations depending on the sign of $b$.

Finally, if $B\neq 0$, then the set of situations will be a three-linked zigzag. For $B>0$ this zigzag is presented in Figure 23 and for $B<0$ in Figure 24.

The actual determination of the sets of admissible situations for each of the players describes at the same time the set of all equilibrium situations in the game.

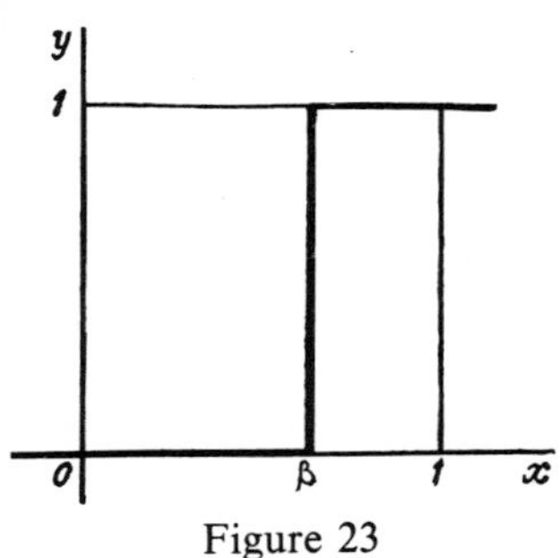

Figure 23

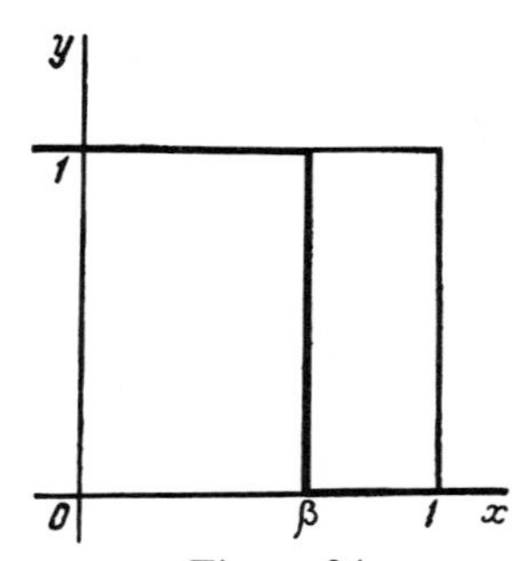

Figure 24

*3.6.7.* We draw attention to the following fact: An admissible situation for a player in a game (cf. Section 1.2.1) depends on the payoff for this player only. Therefore, in our case as well, the sets of all admissible situations for player I depend only on the parameters $A$ and $a$ of his payoff matrix **A**, and the set of all admissible situations for player II only on the parameters $B$ and $b$ determined by the payoff matrix **B** for this player.

In particular, the matrix **A** fully determines the number $\alpha$ that indicates the position of the middle link in the zigzag of admissible situations for player I. In the game-theoretic context, $\alpha$ represents the probability with which player II chooses his or her first pure strategy in a certain mixed strategy for this player. In other words, if $\alpha$ falls into the interval $(0,1)$ it may be interpreted as a mixed strategy for player II. Analogously, the payoff matrix **B** for player II completely determines the number $\beta$, which —if located in the interval $(0,1)$—may be viewed as a mixed strategy for player I.

Assume now that the game possesses an equilibrium situation in completely mixed strategies (i.e. not in pure strategies). It then follows from the above that the mixed equilibrium strategy for player I in this game is completely determined by the payoff matrix for player II and the mixed equilibrium strategy for player II solely by the payoff matrix for player I.

Thus if one of the players in a bi-matrix $2\times 2$ game is "playing" his or her mixed equilibrium strategy and—being certain that the payoffs for his opponent remain unchanged in all situations—decides to reevaluate the situations from his or her own aspect, such a reevaluation should not result in a change of his or her own mixed equilibrium strategy.

*3.6.8.* We now write the explicit expression for $\alpha$ and $\beta$. From (6.10), (6.11), and the definition of $A$, $a$, $B$, and $b$ we have

$$\alpha=\frac{a_{22}-a_{12}}{a_{11}-a_{12}-a_{21}+a_{22}},$$

$$\beta=\frac{b_{22}-b_{21}}{b_{11}-b_{12}-b_{21}+b_{22}}. \tag{6.12}$$

A comparison of these expressions with formula (17.12) in Chapter 1 reveals that in a $2\times 2$ bi-matrix game, under the conditions of the equilibrium situation in completely mixed strategies, the behavior of player II coincides with the behavior of player II in the matrix game with the payoff matrix **A**, while the behavior of player I coincides with the behavior of player II in the matrix game with the payoff matrix **B**.

Consequently, the equilibrium behavior of the players described above is directed towards the minimization of the payoff for the opponent rather than towards the maximization of his or her own payoff.

In other words, "an antagonism in behavior" may arise even without "an antagonism of interests."

## 3.7 Almost antagonistic games

*3.7.1* **Definition.** A bi-matrix game with payoff matrices **A** and **B** is called *almost antagonistic* if the relations $a_{ij}<a_{kl}$ or $a_{ij}=a_{kl}$ imply $b_{ij}>b_{kl}$ or $b_{ij}=b_{kl}$, respectively.

Any matrix game becomes *almost antagonistic* if we use different utility scales in the assessment of the payoffs for each of the players. Typically, player I in an almost antagonistic game is a party that strives to damage his opponent, and that adequately assesses his or her relative damages in various situations but which may in general estimate incorrectly the actual *numerical* values of this damage.

*3.7.2.* We shall now analyze a $2\times 2$ almost antagonistic game.

Without loss of generality (if necessary by transition to strategically equivalent games) we may assume that the payoff matrices in a $2\times 2$

almost antagonistic game are of the form

$$\mathbf{A}=\begin{pmatrix}1 & 0\\ a_{21} & a_{22}\end{pmatrix} \quad\text{and}\quad \mathbf{B}=\begin{pmatrix}-1 & 0\\ b_{21} & b_{22}\end{pmatrix}$$

If $a_{22}<a_{21}$, then by the assumption of almost antagonicity $b_2>b_{21}$ and the second pure strategy for player II dominates his or her first pure strategy. Hence all his or her admissible situations are of the form $(x,0)$. This implies that all the equilibrium situations in the game are either $(1,0)$ or $(0,0)$ or situations of the form $(x,0)$ depending on whether $a_{22}>0$ or $a_{22}<0$ or $a_{22}=0$, respectively.

Now let $a_{22}\geqslant a_{21}$; by assumption this implies that $b_{22}\leqslant b_{21}$. The invariants $A$ and $B$ satisfy in this case:

$$A=1-a_{21}+a_{22}\geqslant 1,$$
$$B=-1-b_{21}+b_{22}\leqslant -1.$$

This shows that $A$ and $B$ are of opposite signs. Checking Figures 21–24, which represent various possible structures of admissible strategies, we observe that the zigzags of the admissible strategies for both players are oriented in the same direction and have therefore only one point of intersection.

*3.7.3.* As an example of an almost antagonistic game, consider the following variant of the *market competition game*.

A small firm (player I) attempts to sell a large amount of merchandise on one of the two markets that is controlled by another, larger firm (player II). For this purpose, it undertakes a specific action at one of the markets (for example, it carries out an intensive advertising campaign). Player II, who dominates the market, may attempt to prevent this campaign by taking certain preventive measures. Player I, if he or she does not find resistance from the market, captures it, while if he or she is challenged, loses it. The strategies for the players are the possible choices of the "markets of concentration" by the firms.

Assume that player I has a greater profit advantage to penetrate the first market, but the struggle for this market is more costly. For example, a victory for player I in the first market may yield a double payoff as compared with the second, but a loss of the first market ruins him or her completely (loss of ten units) while as far as player II is concerned, a victory on the first market results merely in the elimination of a competitor (a gain of five units).

The corresponding bi-matrix game can be defined by the following payoff matrices:

$$\mathbf{A}=\begin{pmatrix}-10 & 2\\ 1 & -1\end{pmatrix},\quad \mathbf{B}=\begin{pmatrix}5 & -2\\ -1 & 1\end{pmatrix}.$$

For this game it is easy to calculate the quantities

$$A = -14 < 0,$$
$$a = -3,$$
$$\alpha = a/A = \tfrac{3}{14}.$$

This implies that the admissible situations for player I are all the situations of the form

$$(0,y), \quad \text{for} \quad \tfrac{3}{14} \leqslant y \leqslant 1,$$
$$\left(x, \tfrac{3}{14}\right), \quad \text{for} \quad 0 \leqslant x \leqslant 1,$$
$$(1,y), \quad \text{for} \quad 0 \leqslant y \leqslant \tfrac{3}{14}.$$

The set of all the admissible situations for player I is presented in Figure 25. Furthermore, we have

$$B = 9 > 0,$$
$$b = 2,$$
$$\beta = b/B = \tfrac{2}{9};$$

hence the admissible situations for player II are all those of the form

$$(x,0), \quad \text{for} \quad 0 \leqslant x \leqslant \tfrac{2}{9},$$
$$\left(\tfrac{2}{9}, y\right), \quad \text{for} \quad 0 \leqslant y \leqslant 1,$$
$$(x,1), \quad \text{for} \quad \tfrac{2}{9} \leqslant x \leqslant 1.$$

The set of these situations is indicated by dotted lines in Figure 25.

The zigzags of admissible situations intersect at the single point $(\tfrac{2}{9}, \tfrac{3}{14})$, which is the unique equilibrium situation of the game.

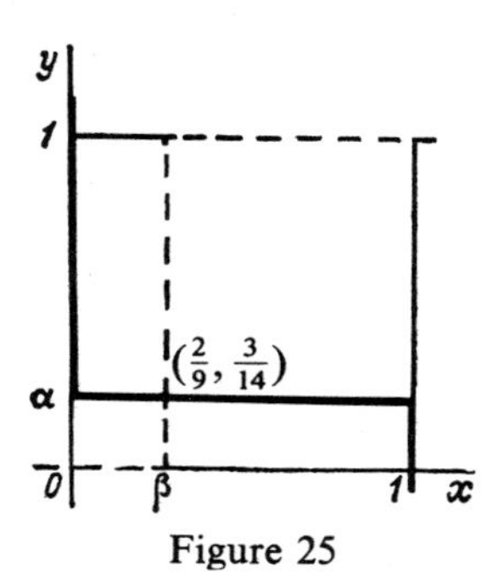


Figure 25

## 3.8 Prisoner's dilemma

*3.8.1.* Assume that players I and II are criminals detained before a trial and are suspected of committing a serious crime. There is no direct evidence against them, however, and their conviction depends heavily on whether the prisoners confess or not.

If both of them confess, they will receive a long-term sentence, but the confession will have a mitigating effect (the losses of each of the players in this case are estimated to be $-8$). If, however, both do not confess, they will be found not guilty of committing the serious crime, but the prosecutor will be able to prove them guilty of a lesser crime and both will be punished appropriately (losses in this case are estimated to be $-1$ for each of them). If, however, only one of the prisoners confesses, then in accordance with the laws of the state, he or she will be set free (the loss is 0) but his or her stubborn partner will be punished extremely severely (the loss estimated to be $-10$).

We shall see that the phenomenon of "positive" affect of antagonicity—detected in the previous example—is even more noticeable in this example.

*3.8.2.* The payoff matrices of the game are

$$\mathbf{A}=\begin{pmatrix}-8 & 0\\ -10 & -1\end{pmatrix},\qquad \mathbf{B}=\begin{pmatrix}-8 & -10\\ 0 & -1\end{pmatrix}.$$

Consequently,

$$A=1>0,$$
$$a=-1,$$
$$\alpha=-1$$

Therefore, the admissible situations for player I are situations of the form $(1,y)$ for any $y$.

Analogously, admissible situations for player II are situations of the form $(x,1)$ for any $x$.

The only equilibrium situation for this game is therefore the situation $(1,1)$ according to which each player should confess. In this situation (see Figure 26), each player suffers a loss of 8 units.

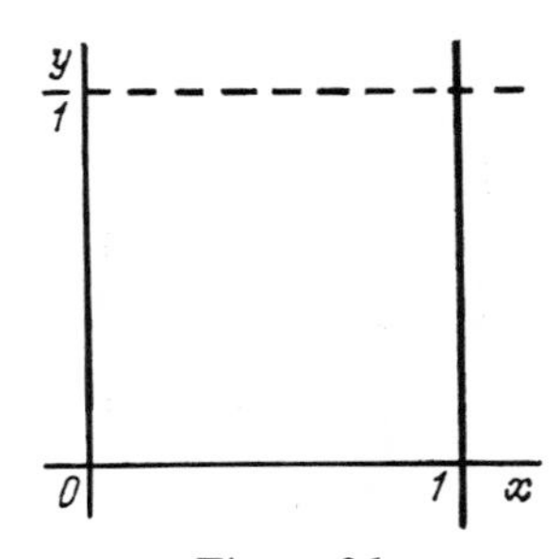


Figure 26

On the other hand, it is evident that situation $(0,0)$, where each player chooses a second pure strategy and the losses for each of the players are minimal (only 1 unit), is very unstable: if one of the players in the situation changes his or her own strategy arbitrarily then that player's payoff increases.

*3.8.3.* This contradiction between the feasibility of a situation expressed in terms of an equilibrium and its "purposefulness," which results in larger payoffs for the players, is basically of the same nature as the contradiction between the maximin and minimax payoffs. Therefore, these contradictions should be resolved analogously by enlarging (extending) the sets of strategies already available. A detailed study of this interesting problem is beyond the scope of our volume.*

## 3.9 The battle of the sexes

*3.9.1.* Two business partners (players I and II) agree on the joint performance of one of two actions: $D_1$ or $D_2$; each of these actions requires joint participation of both partners.

If action $D_1$ is carried out, player I receives one unit of utility and player II receives two units. Conversely, if action $D_2$ is jointly carried out, player I receives two units, while player II just one unit. Finally, if the players each perform different actions, then the payoff for each one equals 0. Thus we have a $2\times2$ bi-matrix game with payoff matrices

$$\mathbf{A}=\begin{pmatrix}1 & 0\\ 0 & 2\end{pmatrix}, \qquad \mathbf{B}=\begin{pmatrix}2 & 0\\ 0 & 1\end{pmatrix}.$$

In various elementary books on game theory, this game is interpreted as a simultaneous choice of the type of an evening's entertainment by a (married) couple: For example, the choices may be a baseball game or a musical comedy: the husband seems to be interested in baseball, while the wife prefers a musical. If the disagreement is not straightened out, then the evening is spoiled. In view of this interpretation the game is often called "the battle of the sexes."

*3.9.2.* Solving this bi-matrix game by applying the theory presented in Section 3.6, we have

$$A=3>0, \qquad a=2, \qquad \alpha=\tfrac{2}{3}.$$

Therefore, situations admissible for player I form a zigzag encompassing the following situations:

$$(0,y), \quad \text{where } 0\leqslant y\leqslant\tfrac{2}{3},$$
$$\left(x,\tfrac{2}{3}\right), \quad \text{where } x \text{ is arbitrary,}$$
$$(1,y), \quad \text{where } \tfrac{2}{3}\leqslant y\leqslant 1.$$

*For a very detailed discussion of this paradox the reader is referred to A. Rappoport's monograph [3E]. An alternative approach to this problem is given in W. W. Hill's paper, Prisoner's dilemma, a stochastic solution, *Math. Magazine* **48** (1975), 103–105 (Translator's remark).

Analogously, we have $B=3$, $b=1$, and $\beta=\frac{1}{3}$. Thus, the admissible situations for player II are

$$\begin{aligned} &(x,0), && \text{where } 0\leqslant x\leqslant\tfrac{1}{3}, \\ &\left(\tfrac{1}{3},y\right), && \text{where } y \text{ is arbitrary} \\ &(x,1), && \text{where } \tfrac{1}{3}\leqslant x\leqslant 1. \end{aligned}$$

We observe from Figure 27 that the game has three equilibrium situations: $(0,0)$, $(1,1)$, and $(\frac{1}{3},\frac{2}{3})$.

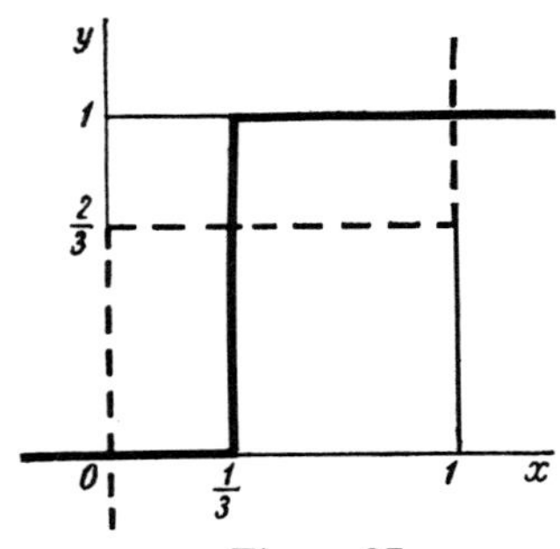


Figure 27

*3.9.3.* The situations $(0,0)$ and $(1,1)$ correspond here to the simultaneous selection by the players of their second or their first pure strategies, respectively, i.e. an agreement to be engaged in a specific joint activity. (Usually any kind of agreement is understood in this sense.)

However, in our case, there is also the third equilibrium situation that consists of choosing certain completely determined *mixed* strategies. Formally this situation may serve as the basis for an agreement to the same extent as the first two. Moreover, this situation seems to be even more "fair" than the first two in the sense that the payoffs for both of the players are the same in this case:

$$\left(\tfrac{1}{3},\tfrac{2}{3}\right)\mathbf{A}\left(\tfrac{2}{3},\tfrac{1}{3}\right)^T=\left(\tfrac{1}{3},\tfrac{2}{3}\right)\mathbf{B}\left(\tfrac{2}{3},\tfrac{1}{3}\right)^T=\tfrac{2}{3}.$$

However, the payoffs for each of the players in this equilibrium situation involving mixed strategies is less than in the case of the first two: we have the payoffs of 1 and 2 for the first situation and 2 and 1 for the second.

We thus observe that the combination of "stability and fairness" contradicts the combination of "stability and profitability."

It is obvious that if both players agreed to "play" their first pure strategy and player II, who benefits from a larger payoff, reimburses player I (with a *side payment*) in the amount of $\frac{1}{2}$, then the payoff in the amount of $1\frac{1}{2}$ should be considered both profitable and fair. However, this type of subdivision of payoff is not studied in the framework of the theory of noncooperative games. They are, however, investigated in the theory of cooperative games, which is discussed in Chapter 4.*

## 3.10 Noncooperative games with two pure strategies for each of the players

*3.10.1.* Simplifications arising in the case when each player possesses only two pure strategies can be utilized not only for bi-matrix games but also in the study of games with an arbitrary finite number of players.

*3.10.2.* Consider a noncooperative game

$$\Gamma = \langle I, \{S_i\}_{i \in I}, \{H_i\}_{i \in I} \rangle$$

in which $S_i = \{1,2\}$ for all $i \in I$. In this case every mixed strategy for player $i$ is completely described by the probability $x_i$ of his or her choosing the first pure strategy. Therefore, the set $\Sigma_i$ of all mixed strategies for the $i$th player can be represented as the segment $[0,1]$ and the set of all the situations in mixed strategies as the unit $n$-dimensional cube. Situations in pure strategies will obviously correspond to the vertices of this cube.

Each vertex of the cube of situations can be represented as an $n$-term sequence of units and twos. In order to identify a vertex in the set of all the vertices, it is sufficient to specify the set of all the players who choose their first (pure) strategy in the situation corresponding to the vertex.

*3.10.3.* We now describe the set of all the admissible situations for player $i$ in the game $\Gamma$.

Choose an arbitrary $K^i \subset 2^{I \setminus i}$ (where $I \setminus i$ is the set $I$ *excluding* $i$) and let $(\alpha_i, K^i)$ denote the situation in which player $i$ chooses his or her pure strategy $\alpha_i$ (here either $\alpha_i = 1$ or $\alpha_i = 2$), players belonging to the set $K^i$ choose their *first* (pure) strategy, and all the remaining players select the second one.

Now let $\sigma$ be an arbitrary situation in mixed strategies and denote by $x_j$, for each player $j \in I \setminus i$, the probability of choosing his or her first strategy. Set

$$\sigma(K^i) = \prod_{j \in K^i} x_j \prod_{\substack{j \notin K^i \\ j \neq i}} (1 - x_j).$$

Obviously,

$$H_i(\sigma) = x_i \sum_{K^i} H_i(1, K^i)\sigma(K^i) + (1 - x_i) \sum_{K^i} H_i(2, K^i)\sigma(K^i). \quad (10.1)$$

*A more detailed description of this game of the battle of the sexes is given in the above quoted monograph by A. Rappoport [3E] (Translator's remark).

If the situation $\sigma$ is admissible for player $i$, then the inequalities $H_i(\sigma \| \alpha_i) \leqslant H_i(\sigma)$ $(\alpha_i = 1,2)$ are satisfied, and conversely.

Substituting the values of the payoff functions $H_i(\sigma\|\alpha_i)$ and $H_i(\sigma)$ as given by (10.1) into the last inequality, we obtain

$$\sum_{K^i} H_i(\alpha_i, K^i)\sigma(K^i)$$

$$\leqslant x_i \sum_{K^i} H_i(1,K^i)\sigma(K^i) + (1-x_i)\sum_{K^i} H_i(2,K^i)\sigma(K^i).$$

Setting successively $\alpha_i = 1$ and $\alpha_i = 2$ and carrying out algebraic simplifications, we have

$$(1-x_i)\sum_{K^i} H_i(1,K^i)\sigma(K^i) \leqslant (1-x_i)\sum_{K^i} H_i(2,K^i)\sigma(K^i), \tag{10.2}$$

$$x_i \sum_{K^i} H_i(1,K^i)\sigma(K^i) \geqslant x_i \sum_{K^i} H_i(2,K^i)\sigma(K^i). \tag{10.3}$$

For $x_i = 0$, the inequality (10.3) is automatically satisfied and (10.2) reduces to

$$\sum_{K^i} H_i(1,K^i)\sigma(K^i) \leqslant \sum_{K^i} H_i(2,K^i)\sigma(K^i). \tag{10.4}$$

For $x_i = 1$ we have the symmetric situation.

*3.10.4.* Denote by $A^i_<$ the set of combinations of strategies for players $I \setminus i$ such that for the situations of the form $(1,\sigma^i)$, where $\sigma^i \in A^i_<$, inequality (10.4) is *strictly* satisfied; by $A^i_=$, the set of combinations of strategies such that for the corresponding situations (10.4) the equality is attained in (10.4); and by $A^i_>$, the set of all the remaining combinations [i.e. such that for the corresponding situations the inequality (10.4) is violated].

It follows from the discussion above that the admissible situations for player $i$ are all the situations of the form

$$(2,\sigma^i), \quad \text{where } \sigma^i \in A^i_<, \qquad (1,\sigma^i), \quad \text{where } \sigma^i \in A^i_>, \tag{10.5}$$

and also all the situations of the form

$$(x_i, \sigma^i) \quad \text{for all} \quad x_i \in [0,1] \quad \text{and} \quad \sigma^i \in A^i_=. \tag{10.5'}$$

## 3.11 False advertising

*3.11.1.* Consider the following conflicting "situation" in the field of advertising.

Three firms (players 1, 2, and 3) put three items (commodities) on the market to cater to a certain class of shoppers that would like to buy this item for Christmas. The shoppers obtain their information about the merchandise from a TV commercial and each firm advertises their own product as the best one. Each firm has two strategies for advertising: (1)

during morning TV programming and (2) during prime time evening TV. If at least two firms advertise their product simultaneously as the best, "sophisticated" shoppers will spot a false claim and will refrain from patronizing these firms. The firm that alone captures the morning audience attains 1 unit of utility in sales and if it alone advertises at night this results in a utility in sales of 2 units.

*3.11.2.* The cube of situations for this game with the payoffs for the players—calculated from the rules of the game—for situations in pure strategies corresponding to the vertices of the cube is presented in Figure 28.

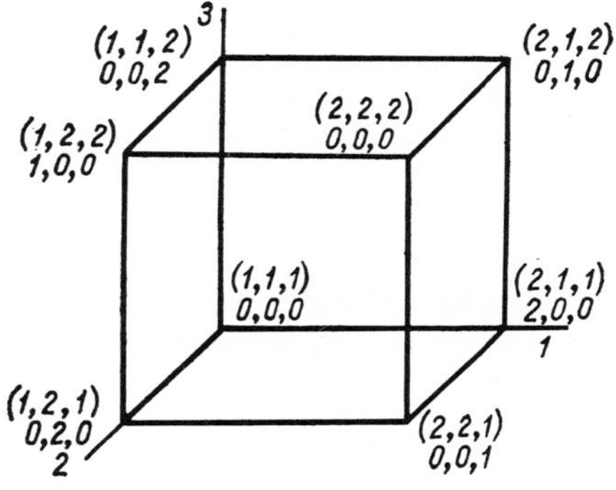

Figure 28

Inequality (10.4) for player I reduces in this case to

$$(1-x_2)(1-x_3)\leqslant 2x_2x_3 \quad \text{or} \quad (1+x_2)(1+x_3)\geqslant 2. \qquad (11.1)$$

The corresponding regions $A^1_<$, $A^1_=$, and $A^1_>$ are presented in Figure 29.

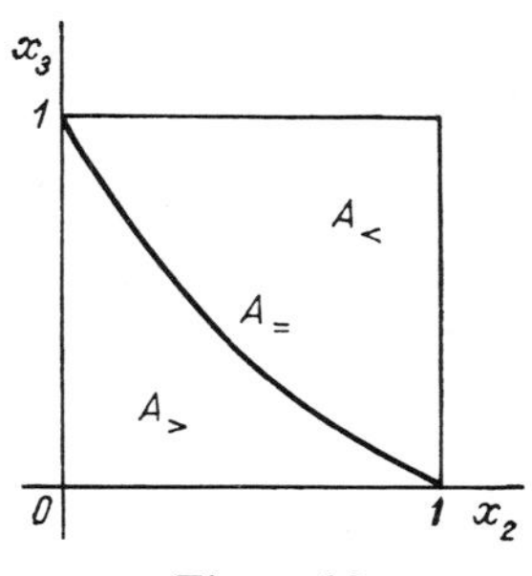

Figure 29

Therefore the set $\mathfrak{S}_1$ of all situations admissible for player I consists of:

(a) situations of the form $(0,x_2,x_3)$, where $(x_2,x_3)\in A^1_<$;

(b) situations of the form $(x_1,x_2,x_3)$, where $(x_2,x_3)\in A^1_=$ and $x_1$ arbitrary; and

(c) situations of the form $(1,x_2,x_3)$, where $(x_2,x_3)\in A^1_>$.

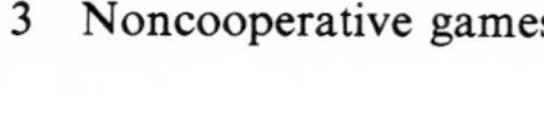



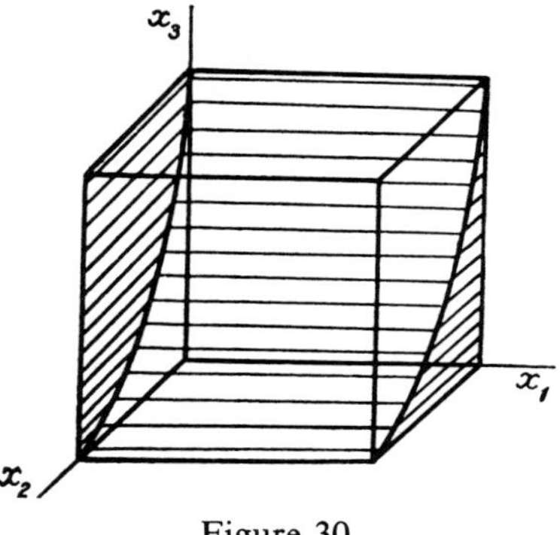


Figure 30

The set of all admissible situations for player I (in the unit cube of all the situations) is shaded in Figure 30. Observe that this set is symmetric with respect to $x_2$ and $x_3$.

*3.11.3.* The sets $\mathfrak{S}_2$ and $\mathfrak{S}_3$ of all the situations admissible for players 2 and 3 is obtained from $\mathfrak{S}_1$ by a corresponding permutation of the coordinates and the set $\mathfrak{S}_\Gamma$ of all the equilibrium situations in this game is determined as the intersection

$$\mathfrak{S}_\Gamma = \mathfrak{S}_1 \cap \mathfrak{S}_2 \cap \mathfrak{S}_3.$$

We now describe this intersection geometrically. First we determine the interior points in the cube belonging to $\mathfrak{S}_\Gamma$. Obviously, these are also the interior points of the cube belonging to each one of the sets $\mathfrak{S}_1$, $\mathfrak{S}_2$, and $\mathfrak{S}_3$. This means that all the coordinates are positive and satisfy the equations

$$(1+x_2)(1+x_3)=2, \qquad (1+x_3)(1+x_1)=2, \qquad (1+x_1)(1+x_2)=2,$$

whence

$$(1+x_1)(1+x_2)(1+x_3)=\sqrt{8} \tag{11.2}$$

and, consequently, $x_1 = x_2 = x_3 = \sqrt{2}-1$ [since all the numbers on the left-hand side of (11.2) are positive, we take the positive root only].

Points of $\mathfrak{S}_1$, $\mathfrak{S}_2$, and $\mathfrak{S}_3$ belonging to the faces of the cube are located correspondingly on the different pairs of opposite faces; the interiors of the different faces of the cube are pairwise disjoint and therefore—in this case—there are no equilibrium situations corresponding to the interior points of the cube.

Finally, each of the sets $\mathfrak{S}_1$, $\mathfrak{S}_2$, and $\mathfrak{S}_3$ contains the very same six edges of the cube together with the vertices located on these edges. Therefore, these six edges are included in the set of equilibrium situations of the game.

The set $\mathfrak{S}_\Gamma$ is presented graphically in Figure 31.

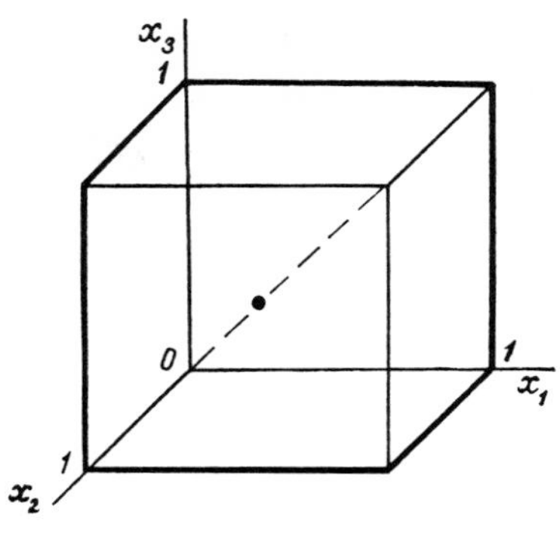


Figure 31

## 3.12 Preservation of ecology

*3.12.1.* Assume that each of three enterprises (players 1, 2, and 3) uses water from a certain natural source (say, a lake) for production purposes. Each enterprise has the following two pure strategies: to construct chemical devices for purification of the sewage water (strategy 1) or to divert it back into the lake without purification (strategy 2). It is also assumed that the source and the technological processes are such that if only one enterprise diverts its sewage back to the lake, this action will not substantially affect the quality of the water, and no loss will be incurred by the enterprises. If, however, at least two users spill the sewage back into the lake, each of the three users will suffer a loss in the amount of two units. The cost of the chemical device for purification of water is estimated as one unit per enterprise.

As in the case of the preceding example, we describe the cube of situations for the game and indicate at each one of the eight vertices of the cube the corresponding payoffs for the three players (Figure 32).

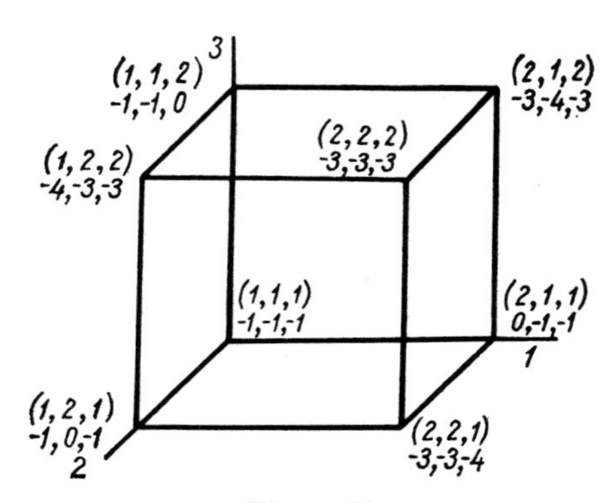


Figure 32

Relation (10.4) for player $i=3$ in this case becomes

$$-x_1x_2-x_1(1-x_2)-(1-x_1)x_2-4(1-x_1)(1-x_2)$$
$$\leqslant -3x_1(1-x_2)-3(1-x_1)x_2-3(1-x_1)(1-x_2), \quad (12.2)$$

or after simple algebraic manipulations,

$$(1-3x_1)(1-3x_2)\geqslant 3x_1x_2. \quad (12.2)$$

The sets $A^3_<$, $A^3_>$, and $A^3_=$ are presented in Figure 33 (here the set $A^3_=$ consists of two curves representing two branches of a hyperbola) and the set of all the admissible situations for player 3 is presented in Figure 34.

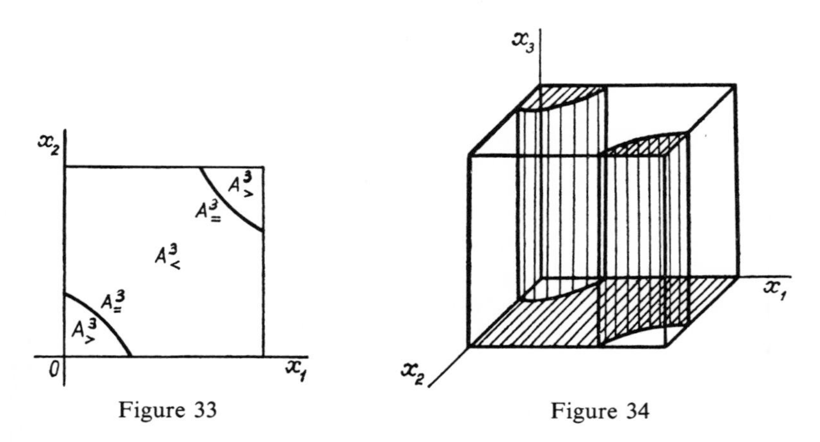


Figure 33

Figure 34

The sets of admissible situations for the other two players are constructed analogously. The intersection of the sets of admissible situations for each of the players gives the set of equilibrium situations of the game. This set consisting of isolated points is indicated in Figure 35.

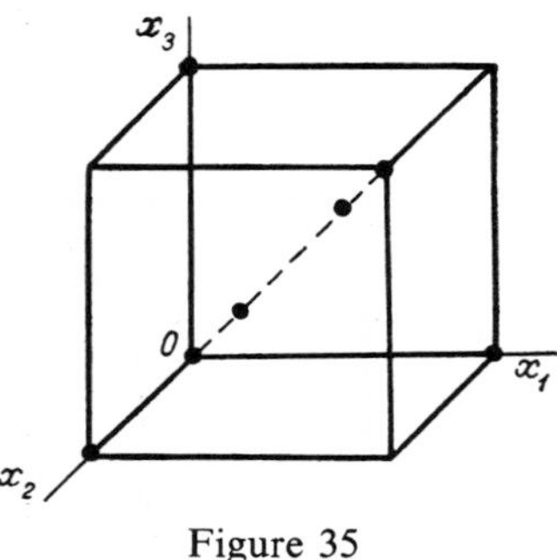


Figure 35

Note that in addition to the ecologically irresponsible equilibrium situation (0, 0, 0) in which all the three enterprises pollute the lake, there exist also two other symmetric equilibrium situations more acceptable from the ecological point of view. These are

$$\left(\frac{1}{3+\sqrt{3}}, \frac{1}{3+\sqrt{3}}, \frac{1}{3+\sqrt{3}}\right) \quad \text{and} \quad \left(\frac{1}{3-\sqrt{3}}, \frac{1}{3-\sqrt{3}}, \frac{1}{3-\sqrt{3}}\right).$$

The latter situation is the most profitable one. In this situation the payoff for each of the players is merely 0.3.

# 4 Cooperative games

## 4.1 Characteristic functions

*4.1.1.* Assume that we have a finite set of players $I$. Any subset $K$ of this set ($K \subset I$) will be called a *coalition*. We shall also consider coalitions consisting of one player and even a void coalition with no players at all. In what follows we shall not distinguish between the *coalition* involving only one player $i$ (singleton $\{i\}$) and the player $i$ and we shall denote this coalition by the symbol $i$. A void coalition will be denoted by the symbol $\varnothing$ (empty set).

Let the players belonging to set $I$ be able—by entering if necessary into relations of exchange and production—to obtain certain comparable payoffs. Denote the payoff that is guaranteed to the coalition $K \subset I$ by $v(K)$. The function $v$ that corresponds the largest guaranteed payoff $v(K)$ to each coalition $K$ is called the *characteristic function*.

Characteristic functions may arise in a great diversity of problems. We shall present a few examples.

*4.1.2* Example. Let $I$ be a group of unskilled workers performing a certain homogeneous task; each of the workers can carry out a certain amount of work and earn (in appropriate monetary units) the amount $a$.

In this case the "profit" of the coalition $K \subset I$ will be $|K|a$ (where, as usual, $|K|$ denotes the number of elements in the set $K$); we thus have $v(K)=|K|a$.

*4.1.3* Example. Assume that in the preceding example each player has special individual skills that increase by the amount $b$ the payoff to his or her co-workers who join him or her in one coalition. In this case the payoff for each of the members of coalition $K$ is equal to $a+(|K|-1)b$ [each member of $K$ utilizes the skills of all the remaining members of the coalition and thus increases his or her payoff by the amount $b(|K|-1)$]. Consequently, the payoff for the whole coalition will be

$$|K|a+|K|\times(|K|-1)b.$$

*4.1.4* Example. Consider a market where the seller (S) offers for sale an indivisible item and prices it in the amount $a$ and two customers $C_1$ and $C_2$ who evaluate this item in the amounts $b$ and $c$, respectively. If $a \geqslant b$, then the seller has no material inclination to make a deal with $C_1$; if $a \geqslant c$, the same is true as far as $C_2$ is concerned. We shall therefore assume that $a < b \leqslant c$.

At the beginning, the distribution of "utilities" among the three participants is

$$a, \quad 0, \quad 0.$$

After the exchange takes place between the seller and customer $C_1$ who purchases the item for the amount $x$, the distribution of "utilities" becomes

$$x, \quad b-x, \quad 0,$$

and after the exchange between the seller and customer $C_2$ who purchases the item for the amount $x$, the utilities are

$$x, \quad 0, \quad c-x.$$

From the above it is easy to verify that in this case the values of the characteristic function for various coalitions are as follows:

$$v(\varnothing)=v(C_1)=v(C_1,C_2)=v(C_2)=0,$$
$$v(S)=a, \quad v(S,C_1)=b,$$
$$v(S,C_2)=v(S,C_1,C_2)=c.$$

## 4.2 Characteristic functions of noncooperative games

*4.2.1.* We now describe a wide (and in a sense universal) source of characteristic functions.

Let a noncooperative game

$$\Gamma=\langle I,\{S_i\}_{i\in I},\{H_i\}_{i\in I}\rangle$$

be given.

Assume that the players who form a coalition $K \subset I$ are united under the conditions of this game for some mutual cause. The question arises, what is the largest payoff that the members of this coalition are guaranteed to obtain jointly?

*4.2.2.* We analyze this question from a game-theoretical point of view. The union of players in $K$ corresponds to a single player, I, say. The fact that the members of the union act jointly means that strategies of the "amalgamated" player I are all the possible combinations of strategies of the "component" players belonging to $K$, i.e. the elements of the Cartesian product $S_K = \prod_{i \in K} S_i$. The common interest of players in $K$ means that the payoff for the "amalgamated" player I is the sum of the payoffs for each of the "component" players in $K$:

$$H_{\mathrm{I}}(s) = \sum_{I \in K} H_i(s).$$

We are trying to determine the largest payoff that can be obtained with certainty by the players of coalition $K$. The only thing that can prevent them from obtaining arbitrarily large payoffs are the rules of the game $\Gamma$ and, in particular, the actions of the players who are not in coalition $K$, i.e. the players belonging to $I \setminus K$. In the worst possible case (as far as player I is concerned), the players belonging to $I \setminus K$ may also unite into a coalition representing a collective player II with the set of strategies $S_{I \setminus K} = \prod_{j \in I \setminus K} S_j$ whose objectives are diametrically opposed to the objectives of player I: $H_{\mathrm{II}}(s) = -H_{\mathrm{I}}(s)$.

*4.2.3.* In view of the discussion above the problem of the largest guaranteed payoff for coalition $K$ in game $\Gamma$ reduces to the problem of the largest guaranteed payoff for player I in the *antagonistic* game $\Gamma_K = \langle S_K, S_{I \setminus K}, H_{\mathrm{I}} \rangle$. However, according to the maximin principle adopted for antagonistic games, this payoff is the value of the game $\Gamma_K$. Clearly, the value of the game $\Gamma_K$ depends, in the final analysis, on coalition $K$ (and also, of course, on the noncooperative game $\Gamma$ that is fixed in our arguments) and is therefore a function of this coalition. This function is said to be the *characteristic function of the noncooperative game* $\Gamma$ and is denoted by $v_\Gamma$. We emphasize that a characteristic function of any noncooperative game is defined on the collection of all the subsets of the set of players and is real-valued.

## 4.3 Properties of characteristic functions for noncooperative games

*4.3.1 Personalization.* For any noncooperative game

$$v_\Gamma(\varnothing) = 0.$$

Indeed according to the definition of the game $\Gamma_\varnothing$

$$H_{\mathrm{I}}(s) = \sum_{i \in \varnothing} H_i(s);$$

however, the sum on the right-hand side is void and therefore $H_{\mathrm{I}}(s)$ is identically zero. Therefore, the payoff function for player I in the game $\Gamma_\varnothing$ is identically zero and hence the value of this game is 0. □

*4.3.2 Superadditivity*. For any noncooperative game $\Gamma$

$$v_\Gamma(K \cup L) \geqslant v_\Gamma(K) + v_\Gamma(L), \quad \text{if } K, L \subset I \text{ and } K \cap L = \varnothing. \tag{3.1}$$

To prove this assertion we observe that

$$v_\Gamma(K \cup L) = \max_{\xi_{K \cup L}} \min_{\eta_{I \setminus (K \cup L)}} \sum_{i \in K \cup L} H_i(\xi_{K \cup L}, \eta_{I \setminus (K \cup L)}),$$

where $\xi_{K \cup L}$ denotes the mixed strategies for coalition $K \cup L$, i.e. arbitrary probability measures on $S_{K \cup L}$ and by $\eta_{I \setminus (K \cup L)}$ probability measures on $S_{I \setminus (K \cup L)}$. If we confine ourselves to those probability measures on $S_{K \cup L}$ that are (direct) products of independent distributions $\xi_K$ and $\xi_L$ on the Cartesian product $S_K \times S_L$, the range of maximization will decrease and the value of the maximum on the right-hand side of the above equation can only decrease. Therefore, we have

$$v_\Gamma(K \cup L) \geqslant \max_{\xi_K} \max_{\xi_L} \min_{\eta_{I \setminus (K \cup L)}} \sum_{i \in K \cup L} H_i(\xi_K, \xi_L, \eta_{I \setminus (K \cup L)}).$$

In this inequality, the left-hand side is not less than the largest possible value of the function to the right of the maxima signs. Therefore, each value of this function satisfies

$$v_\Gamma(K \cup L) \geqslant \min_{\eta_{I \setminus (K \cup L)}} \sum_{i \in K \cup L} H_i(\xi_K, \xi_L, \eta_{I \setminus (K \cup L)}),$$

or

$$v_\Gamma(K \cup L) \geqslant \min_{\eta_{I \setminus (K \cup L)}} \left( \sum_{i \in K} H_i(\xi_K, \xi_L, \eta_{I \setminus (K \cup L)}) + \sum_{i \in L} H_i(\xi_K, \xi_L, \eta_{I \setminus (K \cup L)}) \right).$$

Replacing the minimum of the sum on the right-hand side by the sum of minima (this operation can only decrease the corresponding value), we have

$$v_\Gamma(K \cup L) \geqslant \min_{\eta_{I \setminus (K \cup L)}} \sum_{i \in K} H_i(\xi_K, \xi_L, \eta_{I \setminus (K \cup L)}) + \min_{\eta_{I \setminus (K \cup L)}} \sum_{i \in L} H_i(\xi_K, \xi_L, \eta_{I \setminus (K \cap L)}).$$

Carrying out the minimization of the first summand on the right-hand side with respect to $\xi_L$ and that of the second with respect to $\xi_K$ (to simplify the notation we shall denote these "dummy" variables representing distributions on $L$ and $K$ by $\eta_L$ and $\eta_K$), we can only further reduce the function on the right-hand side:

$$v_\Gamma(K \cup L) \geqslant \min_{\eta_L} \min_{\eta_{I \setminus (K \cup L)}} \sum_{i \in K} H_i(\xi_K, \eta_L, \eta_{I \setminus (K \cup L)}) + \min_{\eta_K} \min_{\eta_{I \setminus (K \cup L)}} \sum_{i \in L} H_i(\eta_K, \xi_L, \eta_{I \setminus (K \cup L)}).$$

In the first summand on the right-hand side minimization is carried out with respect to pairs of *independent* measures on the product $S_L \times S_{I\setminus(K\cup L)}$. If we minimize with respect to an *arbitrary* measure on the product $S_{I\setminus K}$ we increase the domain of minimization and therefore can only decrease the value of the minimum. For the same reasons, the minimization with respect to an *arbitrary* measure on the product $S_{I\setminus L}$ (instead of a pair of *independent* measures on $S_K \times S_{I\setminus(K\cup L)}$) can only decrease the value of the second summand on the right-hand side. As a result of these substitutions we have

$$v_\Gamma(K\cup L) \geqslant \min_{\eta_{I\setminus K}} \sum_{i\in K} H_i(\xi_K, \eta_{I\setminus K}) + \min_{\eta_{I\setminus L}} \sum_{i\in L} H_i(\xi_L, \eta_{I\setminus L}).$$

The last inequality is valid for any values of the measures $\xi_K$ in the first summand and $\xi_L$ in the second. Therefore, taking the maximum with respect to these measures would not alter the sign of the inequality:

$$v_\Gamma(K\cup L) \geqslant \max_{\xi_K} \min_{\eta_{I\setminus K}} \sum_{i\in K} H_i(\xi_K, \eta_{I\setminus K})$$

$$+ \max_{\xi_L} \min_{\eta_{I\setminus L}} \sum_{i\in L} H_i(\xi_L, \eta_{I\setminus L}),$$

whence, by the definition of the characteristic function,

$$v_\Gamma(K\cup L) \geqslant v(K) + v(L),$$

and the assertion is verified. □

*4.3.3 Complementarity*. For any noncooperative *constant sum* game Γ

$$v(K) + v(I\setminus K) = v(I).$$

First observe that for a noncooperative constant sum game

$$v_\Gamma(I) = \sum_{i\in I} H_i(s) = c.$$

Now write

$$v_\Gamma(K) = \max_{\xi_K} \min_{\eta_{I\setminus K}} \sum_{i\in K} H_i(\xi_K, \eta_{I\setminus K})$$

$$= \max_{\xi_K} \min_{\eta_{L\setminus K}} \left(c - \sum_{i\in I\setminus K} H_i(\xi_K, \eta_{I\setminus K})\right)$$

$$= c - \min_{\xi_K} \max_{\eta_{I\setminus K}} \sum_{i\in I\setminus K} H_i(\xi_K, \eta_{I\setminus K}) = c - v_\Gamma(I\setminus K),$$

which proves the assertion. □

*4.3.4.* The set of all characteristic functions with the same set of players form a convex cone (cf. Section 1.12.5): if $v'$ and $v''$ are characteristic functions defined over the set of players $I$ and $\lambda', \lambda'' \geqslant 0$, then $\lambda' v' + \lambda'' v''$ is also a characteristic function over the set $I$.

Indeed

$$\lambda' v'(\varnothing) + \lambda'' v''(\varnothing) = 0$$

and if $K, L \subset I$ with $K \cap L = \varnothing$, then

$$\lambda' v'(K \cup L) + \lambda'' v''(K \cup L) \geqslant \lambda' v'(K) + \lambda' v'(L) + \lambda'' v''(K) + \lambda'' v''(L)$$
$$= (\lambda' v'(K) + \lambda'' v''(K)) + (\lambda' v'(L) + \lambda'' v''(L)).$$

The assertion that if $v'$ and $v''$ satisfy the complementarity condition, then so does the linear combination $\lambda' v' + \lambda'' v''$ is verified analogously.

## 4.4 Imputations and cooperative games

*4.4.1.* Assume that the distribution of utilities available to the set of players in $I$ is such that each player $i \in I$ receives the amount $x_i$. This assignment of utilities can thus be described by the payoff vector $x = (x_1, \ldots, x_n)$.

It is clear that in a specific assignment of utilities, the resulting vectors $x = (x_1, \ldots, x_n)$ cannot be arbitrary but must satisfy certain restrictions that follow from the *conditions* of the utility assignment.

*4.4.2.* Consider a characteristic function $v$ defined on the set of players in $I$. The value of this function $v(K)$ is the total amount guaranteed for the players belonging to the coalition $K$.

We now state the conditions under which a payoff vector $x$ may be considered "admissible" in a noncooperative game with a given characteristic function $v$ (as well as in any other cases leading to payoffs determined by this characteristic function), with subsequent possible unrestricted assignments among the players of the payoffs obtained by various coalitions.

First, it is natural to require that for each $i \in I$

$$x_i \geqslant v(i). \tag{4.1}$$

Indeed, otherwise each player as a member of the coalition will receive less than he or she can obtain acting completely independently without being concerned with agreements with the other players. If player $i$ in a particular distribution of payoffs in assignment $x$ is offered less than $v(i)$, he or she is going to refuse to participate in the coalition and thus this distribution of payoffs will not be realized.

Second, it is necessary that

$$\sum_{i \in I} x_i = v(I). \tag{4.2}$$

Indeed if

$$\sum_{i \in I} x_i < v(I),$$

then all of the players in $I$ will get less together than they can receive under the conditions of the characteristic function $v$. Instead, all of the players in $I$ can achieve together amount $v(I)$ and then subdivide this amount in such a manner that each player $i \in I$ will receive more than his or her share $x_i$. Consequently, the distribution $x$ should be considered unprofitable for all the players and therefore inadmissible.

On the other hand, if the inequality

$$\sum_{i \in I} x_i > v(I),$$

is satisfied, it means that the players in $I$ subdivide an amount which exceeds the one which is at their disposal. It thus follows that such a vector $x$ cannot possibly be realized.

Condition (4.1) is called the condition of *individual rationality* (or *blocking process*), while condition (4.2) is the condition of *group rationality* (it is a special case of *Pareto rationality*).

*4.4.3* **Definition.** the vector $x = (x_1, \ldots, x_n)$ with components $x_i$ assigned to players $i \in I$ and satisfying the conditions of individual and group rationalities is called an *imputation* under the given characteristic function $v$.

**Definition.** The system

$$\langle I, v \rangle, \tag{4.3}$$

consisting of a set of players and a characteristic function on this set and a set of imputations satisfying (4.1) and (4.2) under this characteristic function, is called a *classical cooperative game*.

*4.4.4.* **Theorem.** *In order that the vector $x = (x_1, \ldots, x_n)$ be an imputation in a classical cooperative game $\langle I, v \rangle$, it is necessary and sufficient that*

$$x_i = v(i) + a_i \qquad (i \in I),$$

*Moreover,*

$$a_i \geqslant 0 \qquad (i \in I), \qquad \sum_{i \in I} a_i = v(I) - \sum_{i \in I} v(i). \tag{4.4}$$

PROOF. Sufficiency is established by checking that vector $x$ satisfies the conditions of the individual and collective rationality. To prove necessity, set

$$x_i - v(i) = a_i. \tag{4.5}$$

Equation (4.1) implies that $a_i \geqslant 0$. Summing up termwise all the equalities of the form (4.5) and taking (4.2) into account, we obtain (4.4). □

*4.4.5.* Presently, more general cooperative games are investigated in game theory than the classical cooperative games. In this chapter, we shall, however, confine ourselves to the study of classical cooperative games and refer to them simply as *cooperative games*. Since the characteristic function is the basic and defining component of a cooperative game, these games are often called *games in characteristic function form*.

*4.4.6.* From the standpoint of mathematical treatment and the general approach, the theory of cooperative games is similar to the theory of noncooperative games discussed in the preceding chapters. However, one should always keep in mind the following major features that distinguish these two theories.

To begin with, noncooperative games are strategic games in the sense that an outcome (a situation) is formed as a result of the actions of those players who get certain payoffs in this particular situation. On the other hand, an outcome of a cooperative game is an imputation that arises as a result of an agreement among the players rather than as a consequence of their actions.

Therefore, cooperative games are compared with respect to preferability of payoffs and not of the situations, as is the case in noncooperative games. Moreover, the comparison of imputations is not limited to the individual payoff but is more complex in nature. The difference in the nature of preferences results in different optimality criteria applicable to cooperative games. Moreover, optimality principles for cooperative games are quite diverse and are often very complex. A few of them are discussed below.

## 4.5 Essential and inessential games

*4.5.1.* Superadditivity of the characteristic function reflects the property that a "union" of players into a single coalition (and unions of small coalitions into a larger one) is advisable in order to increase the corresponding payoffs. Essentially this condition reflects the reasonableness of the collective approach and aggregation in economics. The weakest form of superadditivity of a characteristic function is addivity that implies equality in equation (3.1):

$$v(K)+v(L)=v(K\cup L)\quad \text{for all } K,L\subset I, K\cap L=\varnothing.$$

Addivity of the characteristic function reflects the fact that the players are not interested in forming coalitions under the conditions of the given game.

*4.5.2.* A useful criterion for additivity of a characteristic function is given by the following theorem.

**Theorem.** *In order that a characteristic function be additive it is necessary and sufficient that equality*

$$\sum_{i\in I} v(i)=v(I) \tag{5.1}$$

*be satisfied.*

PROOF. Necessity of equality (5.1) is directly implied by the additivity assumption.

To prove sufficiency consider two disjoint coalitions $K$ and $L$ and write the sequence of inequalities implied by the superadditivity of function $v$ and by equation (5.1):

$$v(K)+v(L) \leqslant v(K \cup L),$$

$$\sum_{i \in K} v(i) \leqslant v(K),$$

$$\sum_{i \in L} v(i) \leqslant v(L),$$

$$\sum_{i \in I \setminus (K \cup L)} v(i) \leqslant v(I \setminus (K \cup L)),$$

$$v(K \cup L)+v(I \setminus (K \cup L)) \leqslant v(I),$$

$$v(I) \leqslant \sum_{i \in I} v(i).$$

(Actually the last relationship holds with the equality sign, however the weaker statement is also correct.) Observe now that in the totality of the above inequalities the same terms appear to the right and to the left and the directions of the inequalities are also the same. Hence, adding these inequalities termwise, we obtain an identity. Therefore, each of the original inequalities should be an equality, particularly the first one,

$$v(K)+v(L)=v(K \cup L),$$

which proves the sufficiency part of the theorem. □

**Definition.** A cooperative game with an additive characteristic function is called *inessential*. Other cooperative games are called *essential*.

*4.5.3* **Theorem.** *An inessential game has only one imputation. This is the imputation*

$$(v(1),\ldots,v(n)). \tag{5.2}$$

*On the other hand, any essential game with at least two players possesses infinitely many imputations.*

To prove the theorem, represent the imputation in the form given in the theorem in Section 4.4.4:

$$(v(1)+a_1,\ldots,v(n)+a_n), \qquad a_i \geqslant 0 \qquad (i \in I). \tag{5.3}$$

If the game is inessential, equality (5.1) is satisfied; this implies that $a_i=0$ $(i \in I)$ and we obtain the imputation (5.2).

If, however, the game is essential, then

$$v(I)-\sum_{i\in I} v(i)>0,$$

and this positive difference can be subdivided into nonnegative summands $a_i$ in infinitely many ways. □

## 4.6 Strategic equivalence of cooperative games

*4.6.1.* Since there exist a great variety of cooperative games, it is desirable to classify them in such a manner that the games which belong to the same class will possess the same basic properties. This will allow us—instead of dealing with all the games belonging to a certain class—to consider only a single representative game and, moreover, to choose it to be in a certain sense, the simplest.

Examples of such classes are classes of strategically equivalent games.

**Definition.** A cooperative game with the set of players I and characteristic function $v$ is called *strategically equivalent* to a game with the same set of players and characteristic function $v'$ if a positive number $k$ and arbitrary real numbers $c_i$ ($i\in I$) exist such that for any coalition $K\subset I$ equality

$$v'(K)=kv(K)+\sum_{i\in K} c_i \tag{6.1}$$

is satisfied.

Often instead of speaking of the strategic equivalence of cooperative games, we refer equivalently to the strategic equivalence of their characteristic functions.

Clearly, in a game with characteristic function $v$ the payoffs for the players and coalitions of players differ from those in a game with characteristic function $v'$ only in the amounts of "initial capital" $c_i$ or $\Sigma_{i\in K}c_i$, respectively, and by the scale of units determined by the coefficient $k$. However, the same strategic considerations are equally valid for both games. This explains the term "strategic equivalence." In what follows we shall have occasion to observe the formal similarity between strategically equivalent games (cf. Section 1.3) and their characteristic functions.

We shall denote the strategic equivalence between two characteristic functions $v$ and $v'$ by the symbol $v\sim v'$.

*4.6.2.* It is easy to prove the following three properties of strategic equivalence:

(1) *Reflexivity*. Each characteristic function is strategically equivalent to itself:

$$v\sim v.$$

Indeed, to verify this property, it is sufficient to set $k=1$ and $c_i=0$ for $i\in I$ in (6.1).

(2) *Symmetry*. If $v \sim v'$, then $v' \sim v$.
Let (6.1) be valid. Solving this equation for $v(K)$, we obtain

$$v(K) = \frac{1}{k} v'(K) + \sum_{i \in K} \frac{c_i^*}{k}, \qquad (c_i^* = -c_i)$$

or setting $1/k = k'$, $c_i^*/k = c_i'$, we have

$$v(K) = k'v'(K) + \sum_{i \in K} c_i'.$$

To complete the proof one need only observe that $k' > 0$.

(3) *Transitivity*. If $v \sim v'$ and $v' \sim v''$, then $v \sim v''$.

Write the equation that expresses the strategic equivalence between $v'$ and $v''$:

$$v''(K) = k'v'(K) + \sum_{i \in K} c_i' \qquad (k' > 0).$$

Substituting the expression for $v'(K)$ from (6.1) into this equation, we obtain

$$v''(K) = kk'v(K) + \sum_{i \in K} (c_i' + k'c_i),$$

or setting $kk' = k''$, $c_i' + k'c_i = c_i''$, we have

$$v''(K) = k''v(K) + \sum_{i \in K} c_i'', \quad \text{where } k'' > 0,$$

which shows that $v \sim v''$.

*4.6.3.* We have thus shown that the strategic equivalence, viewed as a relation between characteristic functions, possesses the properties of reflexivity, symmetry, and transitivity. This implies that the set of all characteristic functions can be uniquely subdivided into mutually disjoint classes. In what follows these classes will be called classes of strategic equivalence.

*4.6.4.* The relation of strategic equivalence between the games (and their characteristic functions) induces strategic equivalence between imputations. Namely, let $v \sim v'$:

$$v'(K) = kv(K) + \sum_{i \in K} c_i,$$

and let $x = (x_1, \ldots, x_n)$ be an imputation under the characteristic function $v$. Consider the vector $x' = (x_1', \ldots, x_n')$, where $x_i' = kx_i + c_i$. For this vector

$$x_i' = kx_i + c_i \geqslant kv(i) + c_i = v'(i),$$

i.e. the condition of individual rationality is satisfied. Also

$$\sum_{i \in I} x_i' = \sum_{i \in I} (kx_i + c_i) = k \sum_{i \in I} x_i + \sum_{i \in I} c_i = kv(I) + \sum_{i \in I} c_i = v'(I),$$

i.e. the condition of group rationality is also satisfied. Therefore, the vector $x'$ is an imputation under conditions of characteristic function $v'$. We say that imputation $x'$ *corresponds* to imputation $x$ under the strategic equivalence $v \sim v'$.

## 4.7 Zero games

*4.7.1* **Definition.** A cooperative game with a set of players $I$ is called a *zero game* if its characteristic function is identically zero.

Zero games reflect the total lack of interest of players in the game.

**Theorem.** *Any inessential game is strategically equivalent to a zero game.*

PROOF. The characteristic function of an inessential games satisfies

$$v(I) = \sum_{i \in I} v(i).$$

Therefore the characteristic function $v'$—strategically equivalent to $v$—defined by

$$v'(K) = v(K) - \sum_{i \in K} v(i) \quad \text{for all } K$$

is a zero characteristic function. [From the definition of strategic equivalence we have here $k = 1$ and $c_i = -v(i)$. Observe that $\Sigma_{i \in K} v(i) = v(K)$.] □

As a corollary we obtain that all the inessential games with a given set of players $I$ constitute a single class of strategic equivalence.

## 4.8 The 0-1 reduced form

*4.8.1* **Definition.** A cooperative game with the characteristic function $v$ is represented in the 0-1 *reduced form* if

$$v(i) = 0 \quad (i \in I), \qquad v(I) = 1.$$

**Theorem.** *Every essential cooperative game is strategically equivalent to one and only one game in the* 0-1 *reduced form.*

(This game is often called the 0-1 reduced form of the initial game.)

PROOF. Let $v$ be the characteristic function of an arbitrary essential game. Consider the system of $n+1$ equations in $n+1$ unknowns $c_1, \ldots, c_n$ and $k$:

$$v'(i) = kv(i) + c_i = 0, \qquad i = 1, \ldots, n, \tag{8.1}$$

$$v'(I) = kv(I) + \sum_{i \in I} c_i = 1. \tag{8.2}$$

Adding the $n$ equations of (8.1) termwise, we obtain

$$k\sum_{i\in I} v(i)+\sum_{i\in I} c_i=0;$$

this together with equation (8.2) yields

$$k\left(v(I)-\sum_{i\in I} v(i)\right)=1.$$

Since the game is assumed to be essential, the coefficient multiplying $k$ is positive. Therefore,

$$k=\frac{1}{v(I)-\sum_{i\in I} v(i)}>0. \tag{8.3}$$

We can now determine the unknown $c_i$:

$$c_i=\frac{v(i)}{v(I)-\sum_{i\in I} v(i)}. \tag{8.4}$$

Equalities (8.3) and (8.4) are necessary corollaries of the system of equations (8.1)–(8.2). Therefore, the characteristic function of the 0-1 reduced form of an essential game is uniquely determined. □

*4.8.2.* It is obvious that the characteristic function in the 0-1 reduced form is nonnegative and monotone.

*4.8.3.* Any vector $x=(x_1,\ldots,x_n)$ whose components satisfy

$$x_i\geqslant 0 \quad (i\in I), \qquad \sum_{i\in I} x_i=I$$

can serve as an imputation in a game in the 0-1 reduced form. Therefore, imputations in such a game can be viewed as points in the fundamental $(n-1)$-dimensional simplex described in terms of their barycentric coordinates.

*4.8.4.* Analogously to the 0-1 reduced forms of cooperative games, one can consider an "$a$-$b$ reduced" form of these games for arbitrary $a$ and $b$ (provided $na\neq b$). Games in the "$a$-$b$ reduced" form are games with the characteristic function $v'$ satisfying

$$v'(i)=a \quad (i\in I), \qquad v'(I)=b.$$

It is easy to show that every essential game admits only one $a$-$b$ reduced form for any $a$ and $b$ (provided $na\neq b$).

In addition to the 0-1 reduced form, games in the $-1$-0 reduced form are often encountered in game theory.

## 4.9 Classification of cooperative games with a small number of players

*4.9.1.* All the arguments in this section pertain to classes of strategic equivalence of cooperative games.

In view of the discussion in Section 4.6 a unique class of strategically equivalent inessential games exists for any set of players $I$. Noting this fact, we shall in what follows consider in view of Section 4.8.1 only essential cooperative games in their 0-1 reduced form.

*4.9.2.* We start with constant sum games. *Every cooperative constant sum game with two players is inessential.*

Indeed, assume that there exists an essential cooperative game with two players and characteristic function $v$. As shown in Section 4.8, such a game is strategically equivalent to a game in the 0-1 reduced form with characteristic function $v'$. In our case this implies that

$$v'(1)=v'(2)=0, \tag{9.1}$$

$$v'(1,2)=1. \tag{9.2}$$

Using the complementarity property (Section 4.3), we obtain from (9.1) and (9.2)

$$v'(2)=v'(1,2)-v'(1)=1-0=1,$$

which contradicts (9.1) and proves the assertion.

*4.9.3.* Consider now three-person cooperative games. All the inessential games of this kind belong, as it was mentioned above, to a single class of strategic equivalence. We shall therefore discuss essential games of this kind and enumerate their 0-1 reduced forms. Let $v$ be the characteristic function of an essential game in the 0-1 reduced form. This implies that

$$v(1)=v(2)=v(3)=0, \qquad v(1,2,3)=1.$$

From the complementation property we have

$$v(1,2)=v(1,2,3)-v(3)=1-0=1$$

and analogously $v(1,3)=v(2,3)=1$.

This determines completely the characteristic function of the game. Thus there are only two classes of strategic equivalence of three-person cooperative games, the class of inessential games, and the class of essential ones.

*4.9.4.* The collection of classes of strategic equivalence for four-person cooperative games is substantially larger.

First, as in the case of any number of players, there is a single class of strategic equivalence for inessential games.

We now enumerate all the 0-1 reduced forms of four-person games.

Let $v$ be such a game. Since the characteristic function is in the 0-1 reduced form, we have

$$v(1)=v(2)=v(3)=v(4)=0,$$
$$v(1,2,3,4)=1.$$

The complementarity property implies

$$v(1,2,3)=v(1,2,3,4)-v(1)=1-0=1$$

and analogously $v(1,2,4)=v(1,3,4)=v(2,3,4)=1$.

To complete the description of the characteristic function of the game it is required to prescribe the values of this function for all the coalitions consisting of two players. There are six such coalitions. In view of the complementarity property, the values of the characteristic function of these coalitions must satisfy

$$v(1,4)=1-v(2,3), \qquad v(1,3)=1-v(2,4), \qquad v(1,2)=1-v(3,4).$$

Otherwise there are no other restrictions on $v$. Therefore, three out of the six values can be chosen arbitrarily. Each of the values must, however, belong to the interval [0, 1] since the value of a characteristic function in a coalition consisting of two players is not less than its value in a coalition consisting of one of them (the latter value is 0) and does not exceed its value on a coalition of three players (which equals 1 in this case). Let

$$v(1,4)=x_1, \qquad v(2,4)=x_2, \qquad v(3,4)=x_3.$$

Then

$$v(2,3)=1-x_1, \qquad v(1,3)=1-x_2, \qquad v(1,2)=1-x_3.$$

If we interpret $x_1$, $x_2$, and $x_3$ as coordinates of points in a three-dimensional Euclidean space, then a point in the unit cube (see Figure 36) corresponds to each class of strategic equivalence of a four-person cooperative game. We thus conclude that the collection of classes of strategic equivalence of essential four-person constant sum cooperative games is an

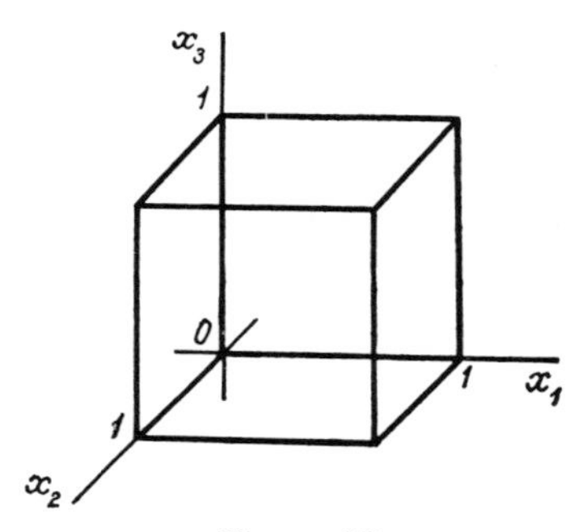

Figure 36

infinite set and is "three-parametric" in the sense that three parameters—values of the characteristic function—are arbitrary.

*4.9.5.* Further investigations reveal that there is an even larger variety of classes of strategic equivalence for constant sum cooperative games with the number of players exceeding four. For example, the dimension of the collection of these classes for five-person games is 10; for six players it is 25, and in general for $n$ players it is $2^{n-1}-n-1$.

*4.9.6.* We now proceed to describe the *nonconstant sum* cooperative games.

We start with the case of $I=\{1,2\}$. Conditions of 0-1 reducibility yield

$$v(\varnothing)=v(1)=v(2)=0,$$
$$v(1,2)=1.$$

Hence there is only a single class of strategic equivalence for essential two-person cooperative games.

*4.9.7.* In the case when $I=\{1,2,3\}$, the condition of 0-1 reducibility implies

$$v(\varnothing)=v(1)=v(2)=v(3)=0,$$
$$v(1,2,3)=1,$$

and the values of the characteristic function

$$v(1,2)=c_3, v(1,3)=c_2, v(2,3)=c_1$$

can be arbitrary numbers in [0, 1]. Thus the classes of strategic equivalence of general three-player cooperative games correspond to the points of the unit cube (as in the case of four-person constant sum games).

## 4.10 Dominance of imputations

*4.10.1.* Let $x=(x_1,\ldots,x_n)$ and $y=(y_1,\ldots,y_n)$ be two imputations in a cooperative game $\Gamma=\langle I,v\rangle$ and let $K\subset I$ be a coalition in this game.

**Definition.** Imputation $x$ *dominates* imputation $y$ *with respect to* (or *through*) coalition $K$ if

$$(1)\quad \sum_{i\in K} x_i \leqslant v(K),$$

(the *effectiveness* condition)

and

$$(2)\quad x_i > y_i \quad \text{for all } i\in K \tag{10.1}$$

(the *preferability* condition).

The dominance relation through a coalition $K$ is denoted by $x \succ_K y$ (or sometimes by $xR_K y$).

Domination is through a coalition $K$, i.e. the relation $x \succ_K y$ expresses a certain preference attributed to $x$ over $y$ by the coalition $K$. The effectiveness condition means that the preferred imputation $x$ should first of all be feasible for this coalition: the sum of the payoff obtained by each of the players belonging to the coalition cannot exceed the amount guaranteed to this coalition as a whole. Otherwise, this coalition, having encountered an imputation that yields more than it can achieve on its own, must accept this imputation unconditionally and should not compare it with any other imputation.

The preferability condition reflects the necessity of a *unanimous* preference towards $x$ by *all* the members of the coalition: If (10.1) is violated for at least one $i$ from $K$, i.e. the imputation $y$ yields a payoff to at least one member of $K$ not smaller than imputation $x$ offers him or her then we can only speak about preferences towards $x$ over $y$ by *some* members of the coalition (those for which 10.1) holds) and not by the coalition as a whole.

*4.10.2* **Definition.** Imputation $x$ *dominates* imputation $y$ if coalition $K$ exists such that $x \succ_K y$.

We denote this relation as $x \succ y$ (or $xRy$). Dominance of $x$ over $y$ signifies that there are "forces" (i.e. the coalition $K$) in "society" (i.e. the totality of players $I$) that prefer $x$ over $y$.

The dominance relation, in general, does not possess properties common to many other relations that would simplify the analysis of the relation. For example, the strict inequality in (10.1) implies that dominance is not a reflexive relation. It may be symmetric (i.e. $x$ may dominate $y$ through a certain coalition, while $y$ dominates $x$ through some other coalition), but not necessarily so. The same remark applies as far as transitivity is concerned. The fact that $x$ dominates $y$ (i.e. there exists a coalition which prefers $x$ over $y$) and $y$ dominates $z$ (i.e. there exists a coalition that prefers $y$ over $z$) does not necessarily imply that $z$ dominates $x$ (i.e., the existence of a coalition that would prefer $x$ over $z$). An example is presented in the Exercises to this section (p. 170).

*4.10.3.* In any given game there are certain coalitions through which dominance cannot possibly be achieved. For example, dominance through a coalition consisting of a single player or through the coalition consisting of all the players in $I$ cannot take place.

Indeed, $x \succ_i y$ would imply that $y_i < x_i \leqslant v(i)$; this however contradicts the individual rationality of the imputation $y$.

Next $x \succ_I y$ would imply $x_i > y_i$ for all $i \in I$, hence

$$\sum_{i \in I} x_i > \sum_{i \in I} y_i = v(I),$$

which contradicts the group rationality of the imputation $x$.

*4.10.4.* The relation of dominance through a coalition (and thus the general relation of dominance) are invariant with respect to strategic equivalence.

**Theorem.** *If $v$ and $v'$ are two strategically equivalent characteristic functions and the imputations $x$ and $y$ correspond respectively to the imputations $x'$ and $y'$ under this equivalence (cf. Section 4.5.4), then $x \succ_K y$ implies $x' \succ_K y'$.*

PROOF. Set

$$v'(K) = kv(K) + \sum_{i \in K} c_i \qquad (k > 0, K \subset I).$$

Dominance $x \succ_K y$ means that

$$\sum_{i \in K} x_i \leqslant v(K) \text{ and } y_i < x_i \quad \text{for } i \in I.$$

However,

$$\sum_{i \in K} x_i' = \sum_{i \in K} (kx_i + c_i)$$
$$= k \sum_{i \in K} x_i + \sum_{i \in K} c_i \leqslant kv(K) + \sum_{i \in K} c_i = v'(K)$$

and

$$y_i' = ky_i + c_i < kx_i + c_i = x_i',$$

i.e. $x' \succ_K y'$. □

*4.10.5.* This theorem indicates that all the phenomena described in terms of dominance of imputations pertain to the classes of strategic equivalence of cooperative games rather than to the games themselves.

In particular, it is sufficient to study these classes for inessential games choosing the zero game as the class representative and the games in the 0-1 reduced form as representatives of various essential games.

*4.10.6.* According to Section 4.5.3 any inessential game possesses only one imputation; therefore no dominations can possibly occur for such games.

*4.10.7.* We shall now describe dominations of imputations in a three-person constant sum essential game (in accordance with Sections 4.6.4 and 4.8.1 it can be assumed that the game is in its 0-1 reduced form).

Let $x = (x_1, x_2, x_3)$ and $y = (y_1, y_2, y_3)$ be two imputations. As it was stated in Section 4.10.3, domination is not allowable through one-member coalitions 1, 2, or 3; neither is it allowable through the three-member coalition $\{1,2,3\}$. In other words, domination here is possible only through one of the two-member coalitions $\{1,2\}$, $\{1,3\}$, or $\{2,3\}$.

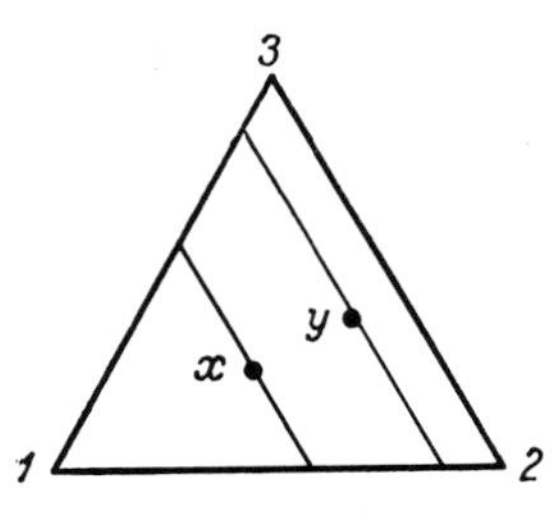

Figure 37

We represent imputations in such a cooperative game as triples of barycentric coordinates of points in a triangle (Figure 37). Domination $x \succ_{1,2} y$ implies that $x_1 + x_2 \leqslant v(1,2)$ and that $y_1 < x_1$ and $y_2 < x_2$. The first of these conditions is automatically satisfied: $x_1 + x_2 = 1 - x_3 \leqslant 1$, while $v(1,2) = 1$. The first part of the second condition $y_1 < x_1$ implies that the straight line parallel to the side *23* of the triangle (Figure 37) that passes through point $x$ is located closer to the vertex *1* of the triangle than the corresponding parallel line that passes through point $y$. Analogously, $y_2 < x_2$ implies that point $x$ is closer to vertex *2* (in the same sense) than point $y$.

Therefore, the totality of all the imputations dominated by the given imputation $x$ through coalition $\{1,2\}$ form an open parallelogram in the triangle of all the imputations. The parallelogram is open in the triangle but not so in the whole plane [in the sense that the sides of the parallelogram located on the sides of the basic triangle belong to the parallelogram, but the sides located inside the triangle do not, since the inequalities in (10.1) are strict (Figure 38)].

Analogously, the set of imputations dominated by the imputation $x$ through the coalitions $\{1,3\}$ and $\{2,3\}$ form another two open parallelograms directed towards the vertices *2* and *3*. Consequently, the set of all the imputations dominated by imputation $x$ form the shaded region presented in Figure 39 in the basic triangle.

It follows directly from the description of the set of all the imputations dominated by an imputation that given two imputations, the necessary and sufficient condition that neither dominates the other, is that the straight

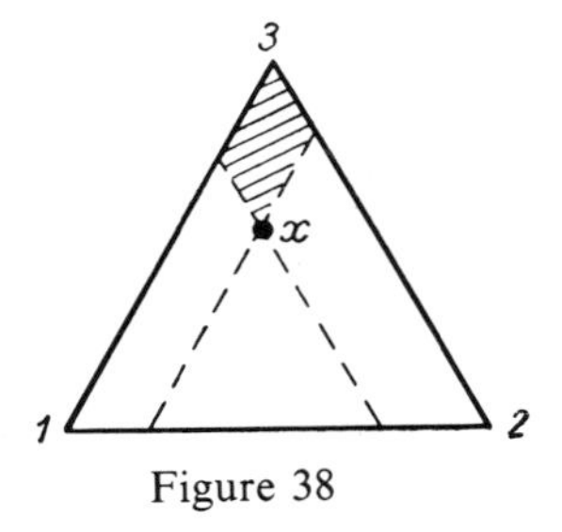

Figure 38

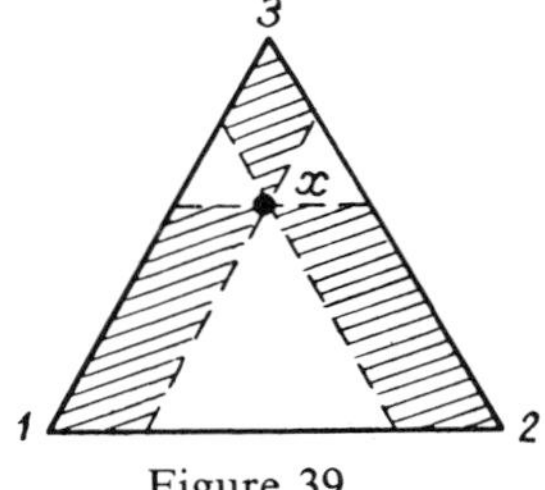

Figure 39

line that passes through the points corresponding to these imputations be parallel to one of the sides of the basic triangle.

*4.10.8.* Consider now the domination of imputations in a general (not necessarily constant sum) three-person game.

We again write explicitly the dominance conditions corresponding to $x \succ_{1,2} y$, where $x=(x_1,x_2,x_3)$ and $y=(y_1,y_2,y_3)$:

$$x_1+x_2 \leqslant v(1,2)=c_3,$$
$$y_1<x_1, \qquad y_2<x_2.$$

Since in general $c_3<1$, the first condition may not be satisfied automatically as it was in the case of constant sum games and therefore cannot be ignored. This condition means that the point in the basic triangle corresponding to the imputation $x$ should be located *not below* the straight line given by equation

$$x_1+x_2=c_3,$$

or, equivalently (recall that we are dealing with barycentric coordinates satisfying $x_1+x_2+x_3=1$), by the equation

$$x_3=1-c_3$$

(see Figure 40). Hence if the imputation $x$ satisfies

$$x_1 \geqslant 1-c_1, \qquad x_2 \geqslant 1-c_2, \qquad x_3 \geqslant 1-c_3, \tag{10.2}$$

then there are three parallelograms of imputations dominated by $x$ (Figure 41). If one of the inequalities in (10.2) is violated (assume for definiteness the third one), then there are two parallelograms of dominated imputations (Figure 42). Finally, if two of the inequalities in (10.2) are violated (assume for definiteness the last two), then there is only one parallelogram of dominated imputations.

Observe that in this case there are more "possibilities" for *non*domination of one imputation by another as compared with the case of constant sum games. These possibilities can be enumerated by analyzing Figures 41–43 and similar drawings that result in assuming other possible violations of inequalities in (10.2). The great variety of cases arising in these situations clearly demonstrates the combinatorial difficulties involved in

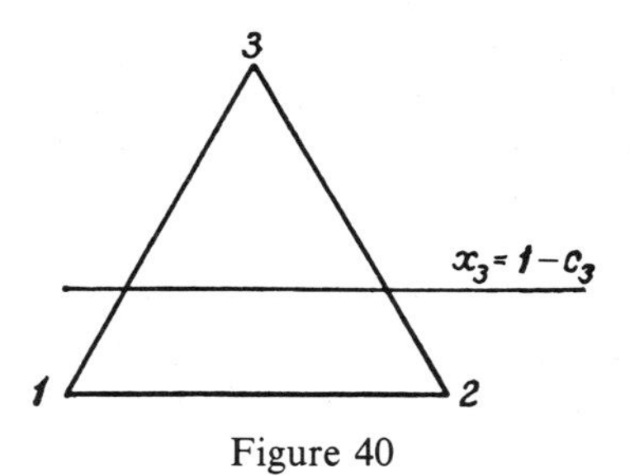


Figure 40

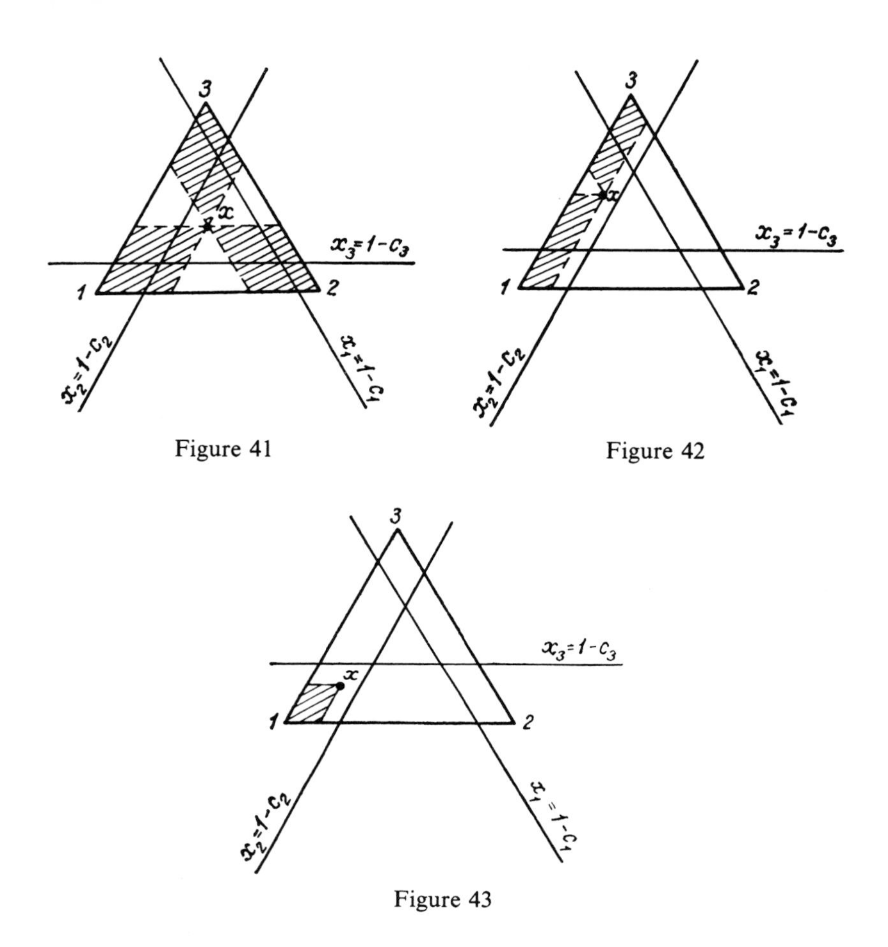


Figure 41

Figure 42

Figure 43

the study of problems connected with domination of imputations in cooperative games as the number of players grows.

## 4.11 The core of a game

*4.11.1.* As it was already mentioned, domination $x \succ y$ may be understood as the preference of imputation $x$ over imputation $y$ by at least one coalition. this in turn means that if someone proposes imputation $y$, the above coalition will come up with a counterproposal suggesting imputation $x$. Therefore an imputation that is not dominated by any other imputation can be considered "fairly stable."

**Definition.** The set of imputations in a coalition game such that no imputation belonging to the set is dominated by some other imputation (i.e. the set of all undominated imputations for a game) is called the *core* of the game.

*4.11.2.* The following theorem presents a convenient characterization of imputations belonging to the core of a game.

**Theorem.** *In order that imputation $x$ belong to the core of a cooperative game with the characteristic function $v$, it is necessary and sufficient that for any coalition $K$ the inequality*

$$v(K) \leqslant \sum_{i\in K} x_i \tag{11.1}$$

*be satisfied.*

PROOF. *Necessity.* In accordance with 4.10.5 it is sufficient to consider games in 0-1 reduced form. For these games the characteristic function is monotone (cf. 4.8.2). Let

$$v(K) > \sum_{i\in K} x_i$$

for the imputation $x$ and for some coalition $K$. Observe that the coalition $K$ must consist of more than one player, otherwise the last inequality violates the individual rationality of imputation $x$. For a similar reason $K$ should be different from $I$. We now have

$$\sum_{i\notin K} x_i = v(I) - \sum_{i\in K} x_i \geqslant v(K) - \sum_{i\in K} x_i > 0.$$

Now choose $\varepsilon$ such that

$$0 < \varepsilon < \frac{1}{|K|}\left(v(K) - \sum_{i\in K} x_i\right),$$

and construct the vector $y=(y_1,\ldots,y_n)$ by setting

$$y_i = \begin{cases} x_i + \varepsilon, & \text{if } i \in K, \\ \dfrac{1}{|I\setminus K|}\left(\sum_{i\notin K} x_i - |K|\varepsilon\right), & \text{if } i \notin K. \end{cases} \tag{11.1'}$$

Direct verification shows that $y$ is an imputation and moreover $y \succ_K x$. Therefore, $x$ does not belong to the core and the necessity of condition (11.1) is verified.

*Sufficiency.* Assuming that imputation $x$ is dominated by imputation $y$ we have for some coalition $K$

$$\sum_{i\in K} x_i < \sum_{i\in K} y_i \leqslant v(K),$$

which violates (11.1) and proves the sufficiency of the condition. □

*4.11.3.* It follows from the preceding theorem that for an imputation to belong to the core of a given cooperative game it is necessary and sufficient that the components of the imputation satisfy a finite system of linear inequalities. This shows that in any cooperative game the core is a (possibly empty) convex polygon.

Clearly the more possibilities of nondomination of one imputation by another in a game, the more likely the game is to possess the core and the larger it will be. The "most favorable" case in this respect is the case of inessential games for which the core exists and consists of the single imputation of this game (see the theorem in Section 4.5.3).

*4.11.4.* The other extreme is the case of a constant sum essential game.

**Theorem.** *For any essential constant sum game the core is empty.*

PROOF. Assume that imputation $x$ belongs to the core. In view of the theorem proved in Section 4.11.2, for each of the players $i$ the inequalities

$$v(i) \leqslant x_i, \tag{11.2}$$

$$v(I \setminus i) \leqslant \sum_{j \neq i} x_j \tag{11.2'}$$

are satisfied; (moreover, in view of Section 4.9.2, the number of players in the game is at least three). Summing up inequalities (11.2) and (11.2′) termwise, we obtain (utilizing the complementarity property and the group rationality of imputations) the equality

$$v(I) = \sum_{i \in I} x_i. \tag{11.3}$$

Therefore the exact equality is attained in (11.2), i.e.

$$v(i) = x_i, \tag{11.4}$$

and this is valid for each $i \in I$. Formulas (11.3) and (11.4) imply the additivity of the characteristic function and thus the *inessentiality* of the game. □

## 4.12 The core of a general three-person game

*4.12.1.* Consider a general three-person cooperative game in its 0-1 reduced form. Its characteristic function satisfies

$$v(\varnothing) = v(1) = v(2) = v(3) = 0, \qquad v(1,2,3) = 1,$$
$$v(1,2) = c_3, \qquad v(1,3) = c_2, \qquad v(2,3) = c_1,$$

where

$$0 \leqslant c_i \leqslant 1 \qquad (i = 1,2,3). \tag{12.1}$$

In view of the theorem in Section 4.11.2, in order that the imputation $x$ belong to the core, it is necessary and sufficient that the inequalities

$$x_1 + x_2 \geqslant c_3, \qquad x_1 + x_3 \geqslant c_2, \qquad x_2 + x_3 \geqslant c_1$$

be satisfied; equivalently,

$$x_3 \leqslant 1 - c_3, \qquad x_2 \leqslant 1 - c_2, \qquad x_1 \leqslant 1 - c_1. \tag{12.2}$$

This implies that the point $x$ is further away from the $i$th vertex of the basic triangle than the straight line

$$\xi_i = 1 - c_i. \tag{12.3}$$

(Note that we are using barycentric coordinates.)

Summing up the inequalities (12.2) termwise, we obtain

$$x_1 + x_2 + x_3 \leqslant 3 - (c_1 + c_2 + c_3),$$

or since the sum of barycentric coordinates is identically unity we have

$$c_1 + c_2 + c_3 \leqslant 2. \tag{12.4}$$

The inequality (12.4) is thus a necessary condition for the existence of a nonempty core in a given game. On the other hand, if (12.4) is satisfied, one can choose nonnegative $\varepsilon_1$, $\varepsilon_2$, and $\varepsilon_3$ such that

$$\sum_{i=1}^{3} (c_i + \varepsilon_i) = 2,$$

and set

$$x_i = 1 - c_i - \varepsilon_i \qquad (i = 1,2,3).$$

These numbers $x_i$ satisfy inequalities (12.2) so that the imputation $x = (x_1, x_2, x_3)$ belongs to the core of the game. Thus condition (12.4) is also sufficient.

*4.12.2.* Geometrically the nonempty core of a cooperative three-person cooperative game is a triangle bounded by the straight lines (12.3) *provided* the intercepts of these lines are situated within the basic simplex of imputations (Figure 44). The latter will hold if the joint solution of *any pair*

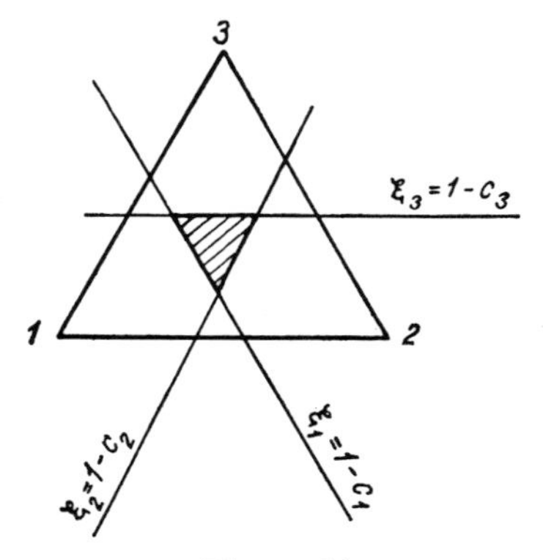


Figure 44

of equations (12.3) and equation

$$\xi_1 + \xi_2 + \xi_3 = 1 \tag{12.5}$$

consists of nonnegative numbers.

Consider for example the system consisting of equations

$$\xi_1 = 1 - c_1, \qquad \xi_2 = 1 - c_2$$

and equation (12.5). [Inequalities (12.1) imply that $\xi_1$ and $\xi_2$ are nonnegative.] The solution of the system yields

$$\xi_3 = 1 - c_1 - c_2,$$

and since $\xi_3 \geqslant 0$ we have

$$c_1 + c_2 \leqslant 1. \tag{12.6}$$

If this condition is not satisfied, then the lower vertex of the shaded triangle falls outside the simplex of imputations (Figure 45) and the core becomes a quadrangle.

A similar role is played by the inequalities

$$c_1 + c_3 \leqslant 1, \qquad c_2 + c_3 \leqslant 1. \tag{12.7}$$

Depending on the possible variations in the sign between the left- and right-hand side of inequalities (12.6) and (12.7) (there are eight possible cases), the core will either be a line segment, triangle, quadrangle, pentagon, hexagon, or a single point. For example, if neither of the inequalities (12.6) and (12.7) is satisfied, the core is a hexagon (see Figure 46).

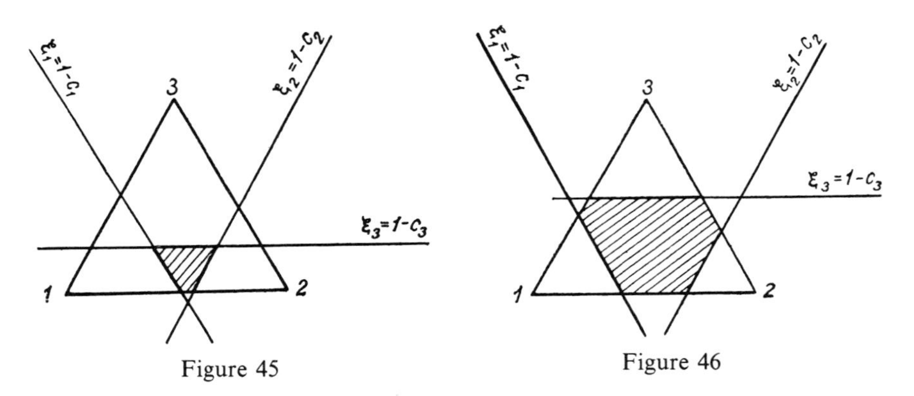


Figure 45 Figure 46

## 4.13 von Neumann–Morgenstern solutions

*4.13.1.* The core, discussed in the preceding section, consists of imputations that are stable but in a somewhat negative or passive sense: the state of affairs does not stimulate us to use an imputation that does not belong to the core. However, the properties of imputations in the core contain no recommendation that these imputations should be used, and moreover these properties do not help us to set them off against other suggested (or recommended) imputations. It would be ideal to have an imputation that, besides being nondominated by any other imputation, in turn dominates any other imputation. Unfortunately, there is no essential cooperative game that possesses such an imputation. It is also impossible to find imputations that possess even a reasonably weakened property of this kind. Therefore, the solution to our problem should be sought along the lines of extending the class of objects to be compared in cooperative games, i.e. along the route of extending the class of imputations. Such a procedure, as have seen before, was fruitful in the case of a noncooperative game: the introduction of mixed strategies there made it possible to solve the problem of the existence of an equilibrium situation for arbitrary finite (and also many infinite) noncooperative games. We shall therefore seek—for cooperative games—a solution in the form of a set of imputations rather than a single "ideal" imputation.

In their classical treatise, J. von Neumann and O. Morgenstern [3A] suggested that a set of imputations that serves as a solution for a cooperative game should satisfy the following two properties: internal stability, which means that imputations belonging to the solution could not "contrast" one another, and external stability, which implies that for any deviation from the solution one can offer a "better" imputation belonging to the solution.

Formally these two properties lead to the following definition:

*4.13.2* **Definition.** The von Neumann–Morgenstern solution (or the vN–M *solution* or the *stable set*) of a cooperative game is called the set $R$ of imputations satisfying the following conditions:

(1) No imputation in $R$ dominates any other imputation in $R$ (internal stability).

(2) If $s$ is an imputation not in $R$, then an imputation $r$ exists belonging to $R$, which dominates $s$ (external stability).

An intuitive interpretation of a vN–M *solution* of a cooperative game is the existence of a system of "behavioral norms" such that no social force (coalition) can contrapose the consequences of two behaviors consistent with these norms, while any deviation from the norm will result in the appearance of some force (i.e. a coalition) in the society (i.e. in the set of all the players $I$) that will strive to restore the norm.

*4.13.3.* A certain relationship between the core and a vN–M solution of a game is given by the following theorem.

**Theorem.** *If a cooperative game possesses both the core $C$ and a* vN–M *solution $R$, then $C \subset R$.*

PROOF. If imputation $x$ belongs to $C$, it cannot be dominated by any other imputation. If it does not belong to the solution, it should be dominated by some imputation belonging to the solution. Consequently, any imputation belonging to the core also belongs to each of the vN–M solutions of the game. □

*4.13.4.* A vN–M solution of a cooperative game cannot consist of a single imputation.

**Theorem.** *If a* vN–M *solution of a cooperative game* $\langle I,v\rangle$ *consists of a single imputation* $x$, *then the characteristic function* $v$ *is inessential.*

PROOF. Assume that the characteristic function $v$ is essential. In view of Section 4.10.5, we may assume that it is in the 0-1 reduced form. Let $x_i$ be a positive component of $x$. In the case of an essential characteristic function $|I|=n>1$, and therefore one can construct an imputation

$$y=(y_1,\dots,y_n)$$

by setting

$$y_j=\begin{cases} x_j+\dfrac{x_i}{n-1}, & \text{if } j\neq i,\\ 0, & \text{if } j=i. \end{cases}$$

By the definition of domination, imputation $x$ does not dominate $y$, hence $y$, being different from $x$, either belongs with $x$ to the vN–M solution of the game, or implies the existence in the vN–M solution of a third imputation $z$ that differs from $x$ (and dominates $y$). □

*4.13.5.* Notwithstanding its numerous advantages, the notion of a vN–M solution possesses certain defects. We shall mention three of them.

First, examples of cooperative games are known for which there is no vN–M solution. Moreover, at the present time, no criteria for the existence of a vN–M solution of a cooperative game are known. Therefore, the optimality principle inherited in a vN–M solution is not universally realized, and the region of its "realizability" so far is undetermined.

Second, a large number of cooperative games possess more than one solution. Hence the optimality principle that leads to a vN–M solution is not complete: in general it does not generate a unique scheme for distributing payoffs.

Finally, as we have seen, solutions for essential cooperative games consist of more than one imputation. Therefore, even if we agree on a particular vN–M solution, we still have not determined (uniquely) the payoff for each of the players.

The defects mentioned above should not be considered as faults to cause revisions of this principle, but as defects that we would like to rectify. Unfortunately, for the games under consideration, this is hardly possible except by drastic reduction of the cases for which this principle is realizable. Moreover, this state of affairs actually reflects the facts of life:

the majority of economic and social problems admit multiple solutions and these solutions are not always directly and consistently comparable with respect to their preferability.

## 4.14 vN–M solutions for three-person constant sum games

*4.14.1.* A vN–M solution for inessential games as well as for arbitrary two-person games is easily determined (as in the case of determination of the core for such games; cf. Section 4.11.3). Namely, there is no dominance of imputations in the case of two-person games, while every inessential game possesses only one imputation. Therefore, the only vN–M solutions for one of these games is the set of all the imputations.

Consequently, a detailed analysis of vN–M solutions of cooperative games should start with the study of three-person essential games. In view of the remarks in Section 4.10.5, we shall confine ourselves to the games in the 0-1 reduced form.

*4.14.2.* The set of imputations for an essential three-person game in its 0-1 reduced form is a triangle. The internal stability of a vN–M solution means no two imputations belonging to the solution dominate each other. Therefore, any two imputations belonging to a single solution should be located on a straight line parallel to one of the sides of the triangle of all the imputations (cf. Section 4.10.7). Consequently, the intervals joining, pairwise, imputations belonging to a vN–M solution should be parallel to the three directions (corresponding to the sides of the triangle of imputations).

We consider separately the following two cases:

(1) The imputations belonging to a vN–M solution are all located on a single straight line.

(2) Not all the imputations belonging to a vN–M solution are situated on a single line.

*4.14.3.* Let the whole vN–M solution $R$ be located on a single line. For definiteness, let the line $AB$ be parallel to the side *12* of the triangle of all the imputations (Figure 47.). No two imputations in $R$ dominate each other so that the internal stability of $R$ is automatically valid.

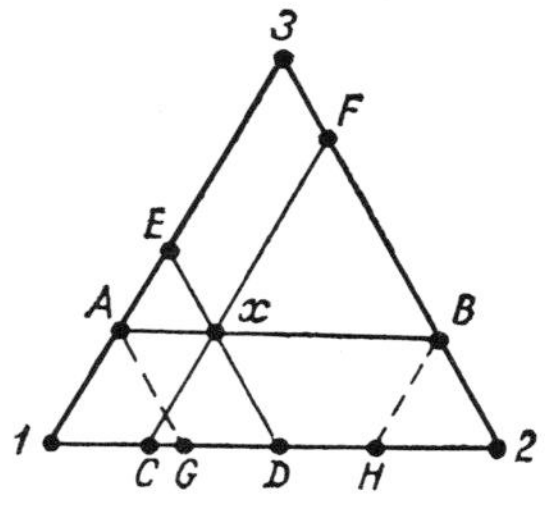


Figure 47

As far as external stability, observe that no imputation on the interval $AB$ is dominated by all the remaining imputations on this interval. Therefore, the external stability of $R$ requires that the interval $AB$ be entirely in $R$. Next, the imputation $x$ (cf. Figure 47) dominates through coalition $\{1,2\}$ all the imputations that form the parallelogram $xF3E$, and all the imputations on this interval dominate through $\{1,2\}$ the union of all these parallelograms, i.e. the triangle $AB3$. Consider now the domination of imputations located below the line $AB$. The imputation $x$ dominates through coalition $\{1,3\}$ the whole parallelogram $xD2B$, and all the imputations on $AB$ the union of such parallelograms, i.e. the parallelogram $AB2G$. Analogously the imputations on $AB$ dominate the parallelogram $BH1A$ through the coalition $\{2,3\}$. Clearly, in order that the imputations on $AB$ dominate all the imputations belonging to the trapezoid $A12B$, it is necessary that the intercept of the lines $AG$ and $BH$ be located strictly below the basis $12$ of the triangle of imputations. We shall express this necessary condition algebraically.

The equation of the line $AB$ is $\xi_3 = x_3$. Hence the points $A$ and $B$ have the barycentric coordinates $(1-x_3, 0, x_3)$ and $(0, 1-x_3, x_3)$, respectively. Therefore, in these coordinates the equations of the lines $AG$ and $BH$ are $\xi_1 = 1-x_3$ and $\xi_2 = 1-x_3$. The condition that the point of intersection of these lines lies outside the triangle $123$ implies that $\xi_1 + \xi_2 > 1$, i.e. $x_3 < \frac{1}{2}$. This means that in order that the set $R$ be externally stable (and therefore constitute a vN–M solution) it is necessary and sufficient that this interval be situated below the median of the triangle. Each interval of this kind represents a vN–M solution of the game under consideration.

The vN–M solutions described above can be "physically" interpreted as follows: Player three obtains a certain fraction $\alpha$ (of the total sum) and the remaining fraction $1-\alpha$ is arbitrarily subdivided between players one and two. We observe that for all imputations in this solution the amount obtained by player three is the same. Such a solution is usually called a *discriminating* solution in the theory of cooperative games and the player who receives the same payoff in all the imputations belonging to this solution is referred to as the *discriminated* one.

Analogously, one can construct the sequences of vN–M solutions of a game in which players one or two, respectively, are *discriminated.*

*4.14.4.* Consider now the case when not all the imputations in a vN–M solution are situated on a straight line.

Let $x$, $y$, and $z$ be three imputations not on the same line. We shall try to "adjoin" to these three imputations a fourth one $u$ that is not dominated by any one of them.

Join $u$ with $x$. The interval $ux$ should be either parallel to one of the sides of the triangle $xyz$ or coincide with it. If the direction of $xu$ coincides with the direction of $xy$ (Figure 48), then the imputation $u$ is on the line $xy$, but differs from both $x$ and $y$. In this case, the interval $zu$ is parallel to neither $zx$ nor $zy$ nor $xy$, which is impossible. Analogously, the case when

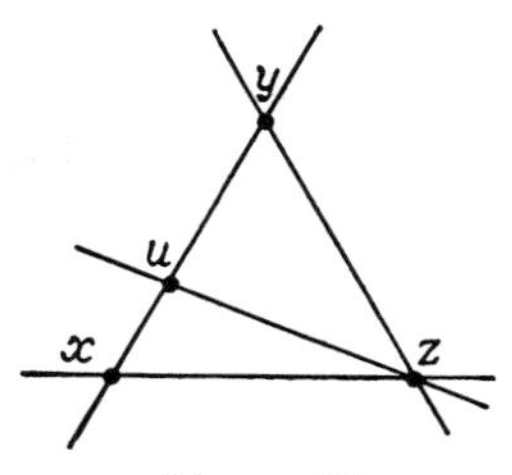

Figure 48

the interval $xu$ is situated along the line $xz$ should be rejected as impossible.

Now, let the interval $xu$ be parallel to $yz$ (Figure 49). In this case, the line joining $z$ with $u$ ought to be parallel to $xy$ and the quadrangle $xyzu$ becomes a parallelogram. The diagonal joining $y$ with $u$ is not parallel to either side of this parallelogram, nor is it parallel to the second diagonal. This yields a fourth direction, which is impossible.

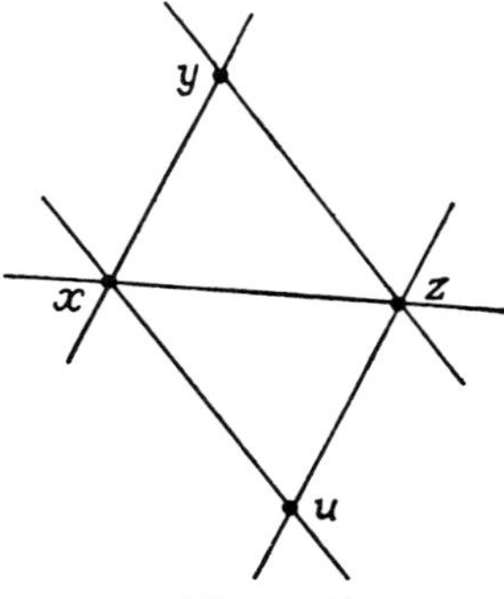

Figure 49

Therefore, if not all the imputations in a vN–M solution are located on a single straight line, then the solution cannot contain four imputations. Therefore it consists of exactly three imputations that form the vertices of a triangle. The sides of this triangle should be parallel to the sides of the triangle of all imputations. Consequently, the small triangle should be situated within the large one in a "similar" (Figure 50) or "anti-similar" fashion (Figure 51).

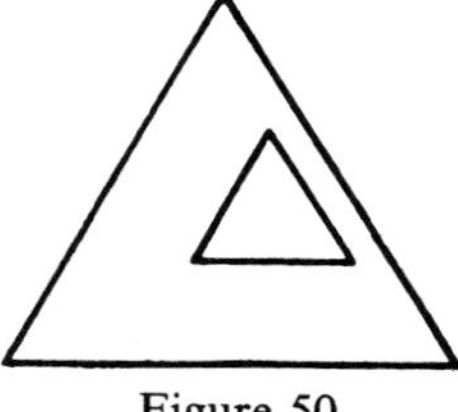
Figure 50

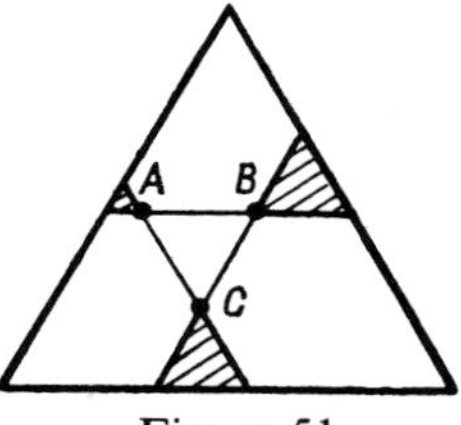

Figure 51

*4.14.5.* Consider first the "similar" case. It is easy to verify that in this case no point of the interior triangle dominates either of its vertices. Hence the triple of the vertices of this triangle cannot form a vN–M solution.

In the second case, as it is easy to verify, all the imputations located in the three closed shaded triangles in Figure 51 remain undominated. Consequently, in order that the set $\{A,B,C\}$ form a vN–M solution, it is necessary that each of these shaded triangles shrink to a single vertex, i.e. that the points $A$, $B$, and $C$ belong to the sides of the basic triangle of all imputations. This is, however, possible only when the points $A$, $B$, and $C$ are located at the middles of the corresponding sides of the basic triangle.

So far our arguments were to establish the necessary conditions for a vN–M solution. We deduced that if a vN–M solution exists in a three-person essential game with imputations not located on the same straight line, then it may consist of only three points, namely, the middles of the sides of the basic triangle (Figure 52). Now we shall verify that the set $\{A,B,C\}$ is indeed a vN–M solution.

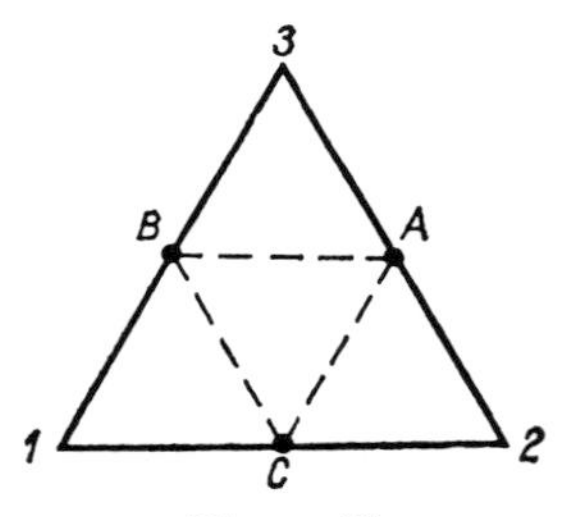


Figure 52

The internal stability of this set up to now was one of the conditions in its construction. Let us check the external stability of the set. Imputation $A$ dominates all the imputations belonging to the parallelogram *1BAC* except for the interior points of the intervals $BA$ and $AC$. Note that it dominates all the interior points of the interval $BC$. Next, imputation $B$ dominates the whole parallelogram *2CBA*, except for the points of the intervals $BC$ and $BA$, while imputation $C$ dominates the parallelogram *CB3A* except for the points of $BC$ and $AC$. However, the interior points of $BC$ are dominated by $A$, those of $AC$ by $B$, and those of $BA$ by $C$. Thus all the imputations except the imputations $A,B,C$ are dominated by these three imputations. Therefore, the set $\{A,B,C\}$ is indeed externally stable and thus it is a vN–M solution.

The unique solution obtained consists of imputations

$$A=\left\{0,\tfrac{1}{2},\tfrac{1}{2}\right\},\qquad B=\left\{\tfrac{1}{2},0,\tfrac{1}{2}\right\},\qquad C=\left\{\tfrac{1}{2},\tfrac{1}{2},0\right\}.$$

It is called a *symmetric* solution.

## 4.15 vN–M solutions for general three-person cooperative games

*4.15.1.* Let us return to the general three-person game considered in Section 4.12. We shall use the notation of that section.

Consider first the case when the *core of the game is empty*. Geometrically it means that the points of pairwise intersections of the straight lines

$$\xi_1 = 1 - c_1, \qquad \xi_2 = 1 - c_2, \qquad \xi_3 = 1 - c_3$$

form a triangle as is shown in Figure 53.

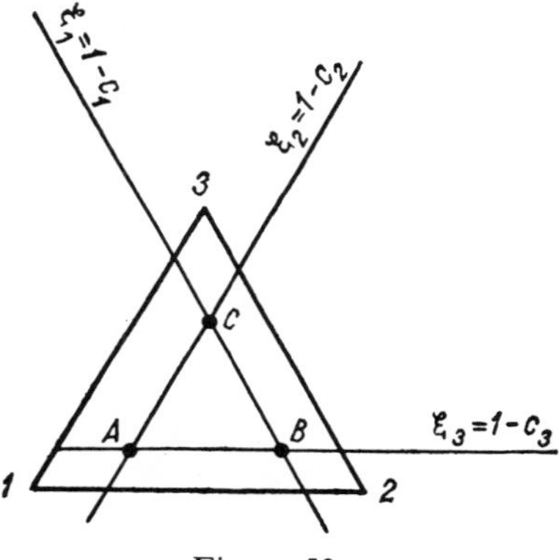


Figure 53

*4.15.2.* It follows from the discussion in Section 4.10.8 that the imputations located within the triangle $ABC$ cannot be dominated by any imputation outside of this triangle. Therefore, every subset of the triangle $ABC$ that is internally and externally stable with respect to domination within the confines of this small triangle will also be stable with respect to domination within the confines of the larger triangle and conversely. For convenience, we shall refer to such a subset of the triangle $ABC$ as "a vN–M solution in small." Thus if a certain vN–M solution exists in the game under consideration, then its intersection with the triangle $ABC$ should be a vN–M solution in small. The latter, however, were described in the previous section.

To obtain a vN–M solution for the whole game from a vN–M solution in small, it is necessary to add certain imputations located outside the triangle $ABC$. We shall consider separately the cases when the solution in small is discriminating or symmetric.

*4.15.3.* Let a vN–M solution in small be discriminating (Figure 54). The set of all imputations not dominated by the set $AB$ is shaded in this figure. We shall restrict ourselves to consideration of a necessary addition to the

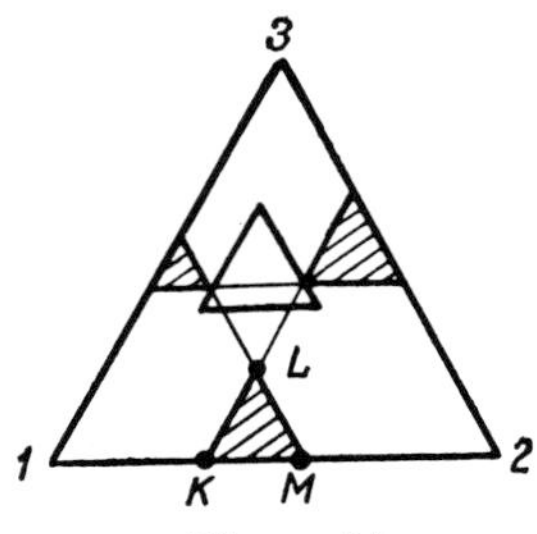


Figure 54

interval $AB$ which will assure that all the imputations in the triangle $KLM$ be dominated. (Additions for the other two triangles are described analogously.) Clearly, such an addition should be located in the triangle $KLM$. The equations of the oblique sides of this triangle are of the form $\xi_1 = \alpha$ and $\xi_2 = \beta$ (for some $\alpha$ and $\beta$). Consider in the triangle $KLM$ a point $x$ with coordinates $(\xi_1, \xi_2)$ (Figure 55). Imputations dominated by this point form two parallelograms (indicated by dotted lines). Consequently, if two imputations belong to the same addition needed to achieve a vN–M solution, then the interval joining them should form at most a 30° angle with the vertical line. In particular, every horizontal line can intersect this addition in at most one point. This means that the whole addition must be situated in a curvilinear interval joining the point $L$ with the basis of the initial triangle of imputations and, moreover, this interval is within 30° of the vertical line. Assume that one of the points of the curvilinear interval does not belong to the addition. In view of its particular location, the point cannot be dominated by any other imputations in the addition. Clearly no imputation outside the addition can dominate this point as well. Therefore, if a point of the curvilinear interval does not belong to the addition, the latter is insufficient to yield a vN–M solution. Thus each point of the curvilinear interval must belong to the addition.

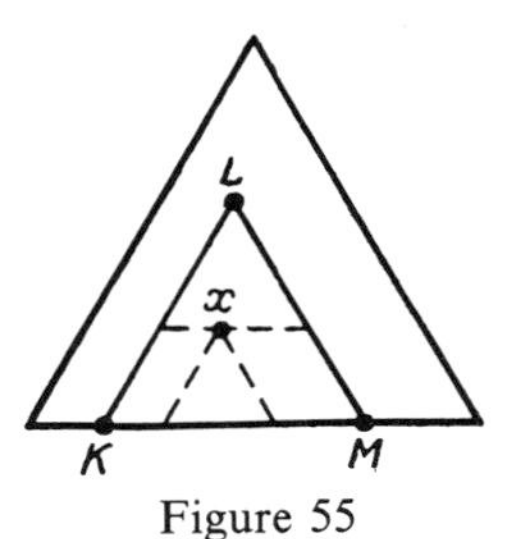


Figure 55

Analogous curves form additions towards a vN–M solution for the two remaining triangles and the whole vN–M solution becomes of the form presented in Figure 56. Clearly, the curves adjoined to a vN–M solution in small should emanate from the corresponding points and deviate from the

Figure 56

perpendicular direction to the corresponding side of the basic triangle of imputations by at most 30°, and finally these curves should go as far as the side of the basic triangle. Otherwise, these additions are completely arbitrary.

*4.15.4.* Now let a vN–M solution in small be symmetric. In this case, one can also supplement it by means of curvilinear intervals as indicated in Figure 57. The restrictions on these curves are the same as those stated in the previous item; otherwise they are arbitrary.

Comparing the discrimination solution with the symmetric one in the general case of a three-person cooperative game, we observe that the latter can be viewed as a limiting case of the first.

Figure 57

*4.15.5.* Finally we discuss the case when the game possesses the *non-empty core*. The collection of imputations that form the core dominates all the imputations except those in the triangles shaded in Figure 58. In order to supplement the core up to a vN–M solution, it is necessary to adjoin to it a curvilinear interval (Figure 59) in each of the triangles of the non-dominated imputations.

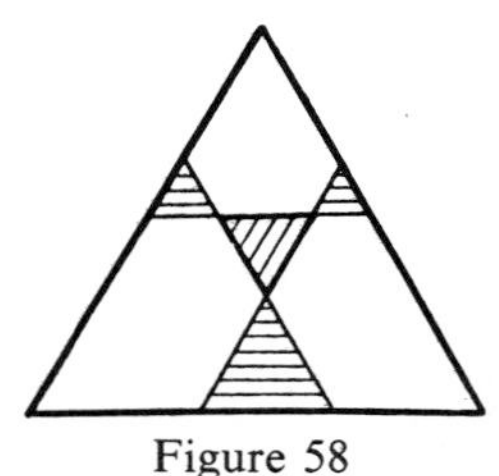

Figure 58

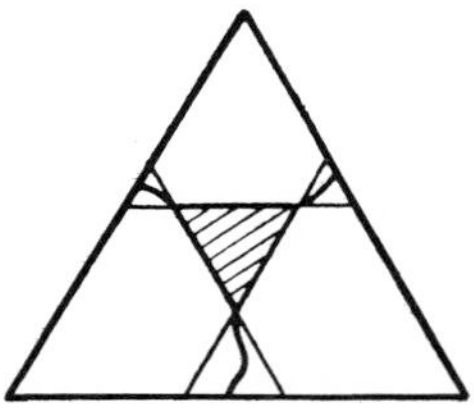

Figure 59

Clearly, if the core is either a quadrangle, pentagon, or hexagon, some of the nondominated triangles will disappear and there will be no need in the corresponding additional curves (Figure 60). In particular, if the core is a hexagon (Figure 46 in Section 4.12), then it coincides with a vN–M solution.

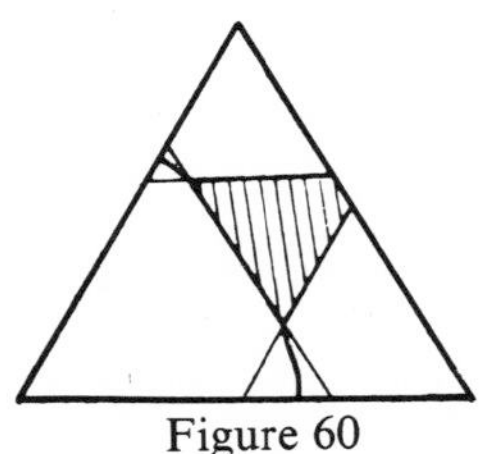

Figure 60

## 4.16 Shapely's vector; axiomatization

*4.16.1.* Game theory studies a variety of optimality principles. Some of these reflect directly our intuitive notions of optimality. Such are, for example, the principle of admissible situations in noncooperative games (cf. Section 1.2)—and its most important particular case the maximin principle—as well as the principles of optimality discussed in the preceding sections that lead to the notions of the core and a vN–M solution. However, in addition to these "natural" principles, game theory derives its own principle of optimality by stipulating the conditions that these principles should satisfy. This is essentially an axiomatic approach to the problem.

In this section we shall consider an axiomatic description of a principle of optimality that is quite interesting from both the theoretical and practical points of view. If we try to delineate the principles of optimality informally, then this principle can be characterized as the *principle* of a *fair subdivision of payoffs*.

Our purpose is to indicate a procedure that corresponds to each cooperative game (i.e. each characteristic function) $v$ over the set of players $I=\{1,2,\dots,n\}$ a vector

$$\Phi(v)=(\Phi_1(v),\Phi_2(v),\dots,\Phi_n(v)),$$

whose components describe fair payoffs in a certain sense, to each of the players participating in the game. It is also desirable that the vector $\Phi(v)$ be an *imputation* under the conditions of the characteristic function $v$.

*4.16.2.* We now state the natural conditions that a fair subdivision of "wealth" should satisfy:

**Definition.** Player $i$, in a cooperative game with characteristic function $v$, is called a *dummy* if for each coalition $K$ not containing $i$,

$$v(K\cup i)=v(K)+v(i). \tag{16.1}$$

In other words, a dummy is a player whose contribution to the coalitions is as large as the amount that he or she can win playing independently. Clearly, if the game is in the 0-1 reduced form, then the inclusion of any number of dummies into a coalition or excluding them from it does not change the value of the characteristic function for this coalition. It is clear from the definition that excluding a dummy from a cooperative game does not affect the game in the least.

**Definition.** The set of players in a cooperative game that consists of all nondummies is called the *support* (*carrier*) *of the game*.

It follows directly from this definition that if $N$ is the support of game $v$, then for any coalition $K$

$$v(K)=v(N\cap K)+\sum_{i\in I\setminus N} v(i).$$

Naturally it is only "fair" to require that when the total payoff for the support of the game is subdivided, no allocation from "community funds" is provided for the dummies and, at the same time, nothing is taken from them.* Formally this statement is expressed by the

**Axiom of effectiveness.** *If $N$ is the support of $v$, then*

$$\sum_{i\in N} \Phi_i(v)=v(N). \tag{16.2}$$

Let $j$ be a dummy in $v$. Consider a support $N$ of $v$ not containing $j$. Clearly the set $N\cup j$ is also a support of $v$. Thus replacing $N$ in (16.2) by $N\cup j$ and applying (16.1) we obtain

$$\sum_{i\in N\cup j} \Phi_i(v)=v(N\cup j)=v(N)+v(j).$$

The last equation together with (16.2) implies that if $j$ is a dummy in $v$, then $\Phi_i(v)=v(j)$.

*4.16.3.* Symmetry is certainly one of the main ingredients of "fairness": players that appear in a game in the same fashion should receive equal payoffs.

**Definition.** Any permutation $\pi$ of players [i.e. a mapping of each player $i$ into $\pi i$ and thus each coalition $K=(i_1,\dots,i_k)$ into coalition $\pi K=(\pi_{i1},\dots,\pi i_k)$] is called an *automorphism* of the characteristic function $v$ if for any coalition $K\subset I$

$$v(\pi K)=v(K).$$

*It should be noted that such a requirement contradicts the idea of a "welfare state" (Translator's remark).

**Axiom of symmetry.** *For any automorphism $\pi$ of the game $v$*

$$\Phi_i(v) = \Phi_{\pi i}(v).$$

*4.16.4.* Consider now two games with the same set of players $I$ and characteristic functions $v'$ and $v''$. The *sum* of these functions, $v' + v''$, defined on all the subsets of $I$ by the relation

$$(v' + v'')(K) = v'(K) + v''(K) \tag{16.3}$$

for each $K \subset I$ possesses all the properties of the characteristic function as was shown in 4.3.4.

The addition of characteristic functions can be viewed as the participation of players in $I$ in two games simultaneously.

It is "fair to assume" that if players participate in two games, their payoffs attained in the individual games should be combined

**Axiom of aggregation.** *If $v'$ and $v''$ are two games, then*

$$\Phi(v' + v'') = \Phi(v') + \Phi(v'').$$

*4.16.5.* It turns out that the system satisfying these three axioms is noncontradictory and complete in the sense that for every characteristic function $v$ there exists a unique vector $\Phi(v)$ satisfying the three axioms.

This vector is called *Shapley's vector* (or *Shapley's value*). In the next section we shall prove the existence and uniqueness of Shapley's vector as well as derive a formula for its actual evaluation.

## 4.17 Shapley's vector; existence and determination

*4.17.1* **Definition.** The characteristic function $v_R$ defined over the set of players $I$ is called *simplest* if

$$v_R(K) = \begin{cases} 1, & \text{if } K \supset R, \\ 0, & \text{if } K \not\supset R. \end{cases}$$

Intuitively the simplest characteristic function describes that state of affairs when a set of players $K$ wins (a "unit" amount) if and only if it contains a certain basic minimal winning coalition $R$.

We shall first determine Shapley's vector for characteristic functions of the form $cv_R$, where $v_R$ is a simplest function and $c$ is a nonnegative number.

Clearly, the set $R$ is the support for $cv_R$ and $v_R(i) = 0$ for $i \notin R$. Hence from the axiom of effectiveness

$$\sum_{i \in R} \Phi_i(cv_R) = cv_R(R) = c.$$

On the other hand the player $i \notin R$ is a dummy; thus

$$\Phi_i(cv_R) = cv_R(i) = 0.$$

Furthermore, it is easy to verify that any permutation of players in which the players in $R$ remain in $R$ is an automorphism of $cv_R$. Hence in view of the axiom of symmetry all the numbers $\Phi_i(cv_R)$ for $i \in R$ are equal to each other; thus

$$\Phi_i(cv_R) = \begin{cases} \dfrac{c}{|R|} & \text{if } i \in R, \\ 0 & \text{if } i \notin R. \end{cases} \tag{17.1}$$

*4.17.2.* We shall now represent arbitrary games in terms of simplest games.

**Lemma.** *For any characteristic function $v$ defined over $I$ there exists a unique linear representation of this function*

$$v = \sum_{R \subset I} \lambda_R v_R \tag{17.2}$$

*in terms of simplest characteristic functions with real coefficients.*

PROOF. Set

$$\lambda_R = \sum_{S \subset R} (-1)^{|R|-|S|} v(S) \tag{17.3}$$

and substitute this expression for $\lambda_R$ into (17.2). For any $T \subset I$ we then obtain

$$v(T) = \sum_{R \subset I} \left( \sum_{S \subset R} (-1)^{|R|-|S|} v(S) \right) v_R(T).$$

Clearly, in this representation of $v(T)$, the only summands that do not vanish are those for which $R \subset T$; moreover, from the definition of $v_R$, $v_R(T) = 1$ for these summands. Therefore,

$$v(T) = \sum_{R \subset T} \sum_{S \subset R} (-1)^{|R|-|S|} v(S),$$

or, interchanging the order of summation,

$$v(T) = \sum_{\substack{S \subset T \\ \times}} \left( \sum_{\substack{S \subset R \subset T \\ \times}} (-1)^{|R|-|S|} \right) v(S)$$

(where the crosses $\times$ indicate the variable sets with respect to which the summation is carried out). Combining all the summands in the interior sum with the same number of players in coalition $R$, we obtain

$$v(T) = \sum_{S \subset T} \left\{ \sum_{r=|S|}^{|T|} \sum_{\substack{|R|=r \\ S \subset R \subset T}} (-1)^{r-|S|} \right\} v(S),$$

and since a set $R$ of $r$ elements containing a given set $S$ of $|S|$ elements can be chosen from a given set $T$ of $|T|$ elements in $\binom{|T|-|S|}{r-|S|}$ ways, we have

$$v(T)=\sum_{S\subset T}\left[\sum_{r=|S|}^{|T|}(-1)^{r-|S|}\binom{|T|-|S|}{r-|S|}\right]v(S).$$

Separating the term that corresponds to $S=T$ and using the binomial formula, we obtain

$$v(T)=v(T)+\sum_{\substack{S\subset T\\ S\neq T}}(1-1)^{|T|-|S|}v(S)=v(T),$$

and the validity of representation (17.2) is verified.

To prove the *uniqueness* of this representation we shall assume that two representations exist of the form (17.2):

$$v(T)=\sum_{R\subset I}\lambda_R v_R(T)=\sum_{R\subset I}\lambda'_R v_R(T). \tag{17.4}$$

Let us prove that $\lambda_R=\lambda'_R$ by induction on $R$ "from all the subsets of the set $R$ to the set itself."

If we set $T=i$ in (17.4), then all the summands on the right- and left-hand side except those corresponding to $R=i$ will vanish and since $v_i(i)=1$, we thus obtain $\lambda_i=\lambda'_i$.

Now choose a coalition $S$ and assume that for all $R\subset S$ $(R\neq S)$ equality $\lambda_R=\lambda'_R$ is valid. If we set $T=S$ in (17.4), then by the induction assumption all the summands corresponding to $R\subset S$ $(R\neq S)$ will be equal while the summands corresponding to $R\not\subset S$ will vanish. This implies that the summands corresponding to $R=S$ should be equal, which proves the uniqueness of the representation. □

*4.17.3.* The coefficients in the representation (17.2) may be of different signs. The following lemma will justify our further manipulations with sums consisting of terms with alternating signs.

**Lemma.** *If the difference $v'-v''$ of two characteristic functions is a characteristic function, then*

$$\Phi(v'-v'')=\Phi(v')-\Phi(v'').$$

PROOF. We have $v'=(v'-v'')+v''$. Therefore, in view of the axiom of aggregation,

$$\Phi(v')=\Phi(v'-v'')+\Phi(v''),$$

which proves the assertion. □

*4.17.4.* We now proceed to compute Shapley's vector for an arbitrary characteristic function $v$. In view of the lemma in 4.17.2, such a function can be represented as a linear combination of simplest functions:

$$v = \sum_{R \subset I} \lambda_R v_R.$$

In this sum we separate summands with nonnegative and negative values of $\lambda_R$ obtaining

$$v = \sum_{\substack{R \subset I \\ \lambda_R \geqslant 0}} \lambda_R v_R + \sum_{\substack{R \subset I \\ \lambda_R < 0}} \lambda_R v_R.$$

Hence,

$$v = \sum_{\substack{R \subset I \\ \lambda_R \geqslant 0}} \lambda_R v_R - \sum_{\substack{R \subset I \\ \lambda_R < 0}} (-\lambda_R) v_R.$$

The sums appearing in the right-hand sides of these equations are characteristic functions being sums of such functions. In view of the condition, their difference $v$ is also a characteristic function. Therefore, from the lemma in 4.17.3 we obtain

$$\Phi_i(v) = \Phi_i\left[\sum_{\substack{R \subset I \\ \lambda_R \geqslant 0}} \lambda_R v_R\right] - \Phi_i\left[\sum_{\substack{R \subset I \\ \lambda_R < 0}} (-\lambda_R) v_R\right],$$

and applying the axiom of aggregation we have

$$\Phi_i(v) = \sum_{\substack{R \subset I \\ \lambda_R \geqslant 0}} \Phi_i(\lambda_R v_R) - \sum_{\substack{R \subset I \\ \lambda_R < 0}} \Phi_i(-\lambda_R v_R).$$

Taking (17.1) into account we thus obtain

$$\Phi_i(v) = \sum_{\substack{i \in R \subset I \\ \lambda_R \geqslant 0}} \frac{\lambda_R}{|R|} + \sum_{\substack{i \in R \subset I \\ \lambda_R < 0}} \frac{\lambda_R}{|R|} = \sum_{i \in R \subset I} \frac{\lambda_R}{|R|}. \tag{17.5}$$

*4.17.5.* It now remains only to replace $\lambda_R$ appearing in (17.5) by its explicit expression given in (17.3) to get

$$\Phi_i(v) = \sum_{i \in R \subset I} \frac{1}{|R|} \sum_{S \subset R} (-1)^{|R|-|S|} v(S) \tag{17.6}$$

and then to simplify the obtained expression as follows:

Interchanging the order of summation in (17.6) yields

$$\Phi_i(v) = \sum_{S \subset I} \left[\sum_{\substack{R \supset S \cup i \\ R \subset I}} \frac{1}{|R|} (-1)^{|R|-|S|}\right] v(S). \tag{17.7}$$

The sum on the right-hand side of (17.7) appearing in parenthesis depends on coalition $S$ and on player $i$. Denote this sum by $\gamma_i(S)$. Clearly if $i \in S$, then

$$\gamma_i(S\setminus i) = \sum_{\substack{R \supset (S\setminus i)\cup i \\ R \subset I}} \frac{1}{|R|}(-1)^{|R|-|S\setminus i|}$$

$$= \sum_{\substack{R \supset S(=S\cup i) \\ R \subset I}} \frac{1}{|R|}(-1)^{|R|-|S|+1} = -\gamma_i(S).$$

Now combining pairwise each coalition $S \subset I$ containing player $i$ with the coalition $S\setminus i$, we obtain

$$\Phi_i(v) = \sum_{\substack{S \subset I \\ S \ni i}} (\gamma_i(S)v(S) + \gamma_i(S\setminus i)v(S\setminus i))$$

$$= \sum_{\substack{S \subset I \\ S \ni i}} \gamma_i(S)(v(S) - v(S\setminus i)).$$

Furthermore, for all such $S$ (containing $i$)

$$\gamma_i(S) = \sum_{\substack{R \supset S \\ R \subset I}} \frac{1}{|R|}(-1)^{|R|-|S|} = \sum_{r=|S|}^{n} \frac{1}{r} \sum_{\substack{|R|=r \\ R \supset S \\ R \subset I}} (-1)^{r-|S|}.$$

The last sum consists of $\binom{n-|S|}{r-|S|}$ equal summands. Consequently,

$$\gamma_i(S) = \sum_{r=|S|}^{n} \frac{1}{r}(-1)^{r-|S|}\binom{n-|S|}{r-|S|}.$$

However

$$\sum_{r=t}^{n} \frac{1}{r}(-1)^{r-t}\binom{n-t}{r-t}$$

$$= \sum_{r=t}^{n} \left(\int_0^1 x^{r-1}dx\right)(-1)^{r-t}\binom{n-t}{r-t}$$

$$= \int_0^1 \sum_{r=t}^{n} x^{r-t}(-1)^{r-t}\binom{n-t}{r-t} x^{t-1}dx = \int_0^1 (1-x)^{n-t}x^{t-1}dx.$$

We now proceed to evaluate the last integral (which is known as the *Beta function*). Denoting the integral by $I(n,t)$ we observe that

$$I(n,1) = \int_0^1 (1-x)^{n-1}dx = -\frac{1}{n}(1-x)^n\Big|_0^1 = \frac{1}{n}. \tag{17.8}$$

We now integrate $I(n,t)$ by parts by setting

$$(1-x)^{n-t}=u, \qquad x^{t-1}dx=dv.$$

Clearly,

$$du=-(n-t)(1-x)^{n-t-1}, \qquad v=\frac{1}{t}x^t.$$

Hence, for $t>0$ we have

$$\begin{aligned} I(n,t) &= \int_0^1 (1-x)^{n-t}x^{t-1}dx \\ &= (1-x)^{n-t}\frac{1}{t}x^t\Big|_0^1 + \frac{n-t}{t}\int_0^1 (1-x)^{n-t-1}x^t dx \\ &= \frac{n-t}{t}I(n,t+1), \end{aligned}$$

which implies the recursive relation

$$I(n,t+1)=\frac{t}{n-t}I(n,t).$$

Applying this relation to (17.8) we obtain successively

$$I(n,2)=\frac{1}{n}\cdot\frac{1}{n-1},$$

$$I(n,3)=\frac{1}{n}\cdot\frac{1}{n-1}\cdot\frac{2}{n-2},$$

and in general

$$I(n,t)=\frac{1}{n}\cdot\frac{1}{n-1}\cdot\frac{2}{n-2}\cdots\frac{t-1}{n-t+1}=\frac{(n-t)!(t-1)!}{n!}.$$

Finally we get the following expression for $\gamma_i(S)$:

$$\gamma_i(S)=\frac{(n-|S|)!(|S|-1)!}{n!},$$

whence,

$$\Phi_i(v)=\sum_{S\ni i}\frac{(n-|S|)!(|S|-1)!}{n!}(v(S)-v(S\setminus i)). \tag{17.9}$$

*4.17.6.* We need only verify that the vector $\Phi(v)$ with components given by (17.9) satisfies the three axioms through which it is derived.

As a preliminary we show that vector $\Phi(v)$ is indeed an imputation.

*4.17.7.* First we shall verify the individual rationality of $\Phi(v)$. In view of superadditivity of the characteristic function we have for $i\in S$

$$v(S)-v(S\setminus i)\geqslant v(i) \tag{17.10}$$

so that

$$\Phi_i(v) \geqslant \sum_{i \in S \subset I} \frac{(n-|S|)!(|S|-1)!}{n!} v(i)$$
$$= v(i) \sum_{i \in S \subset I} \frac{(n-|S|)!(|S|-1)!}{n!}.$$

Combining the terms which correspond to coalitions $S$ with the same number of players we obtain (cf. the argument in 4.17.5)

$$\Phi_i(v) \geqslant v(i) \sum_{s=1}^{n} \sum_{\substack{|S|=s \\ i \in S \subset I}} \frac{(n-s)!(s-1)!}{n!},$$

and since each inner sum consists of $\binom{n-1}{s-1}$ equal terms, we have

$$\Phi_i(v) \geqslant v(i) \sum_{s=1}^{n} \frac{(n-1)!}{(s-1)!(n-s)!} \frac{(n-s)!(s-1)!}{n!} = v(i) \sum_{s=1}^{n} \frac{1}{n} = v(i). \tag{17.11}$$

*4.17.8.* To prove the group rationality of $\Phi(v)$ we write

$$\sum_{i \in I} \Phi_i(v) = \sum_{i \in I} \sum_{S \ni i} \frac{(n-|S|)!(|S|-1)!}{n!} (v(S) - v(S \setminus i)). \tag{17.12}$$

In the double sum on the right-hand side of (17.12) $v(S)$ appears as a minuend $|S|$ times (corresponding to the number of elements contained in $S$) and will thus acquire the coefficient

$$|S| \frac{(n-|S|)!(|S|-1)!}{n!} = \frac{(n-|S|)!|S|!}{n!}$$

which for $S = I$ equals 1. As a subtrahend, the term $v(S)$ appears $n-|S|$ times (corresponding to the number of elements not contained in $S$) and thus will have the coefficient

$$-(n-|S|) \frac{(n-|S|-1)!|S|!}{n!} = -\frac{(n-|S|)!|S|!}{n!} \quad \text{if } n \neq |S|,$$

and coefficient 0 if $n = |S|$, i.e., if $S = I$. Hence the right-hand side of (17.12) becomes $v(I)$ which proves the assertion of the group rationality of $\Phi(v)$.

*4.17.9.* We now proceed to verify the axioms.

1 *Effectiveness*. If player $i$ is a dummy under the conditions of characteristic function $v$, then it follows from the definition that (17.10) is satisfied for this player with the equality sign. Therefore all succeeding relations are valid with the equality sign including relation (17.11).

2 *Symmetry*. Now let $\pi$ be an automorphism of $v$. This means that $v(\pi S) = v(S)$ for any $S \subset I$. As the set $S$ runs through all the coalitions of the game so does $\pi S$. Therefore, on account of (17.9) we can write

$$\Phi_{\pi i}(v) = \sum_{\pi S \subset I} \frac{(n - |\pi S|)!(|\pi S| - 1)!}{n!}(v(\pi S) - v(\pi(S \setminus i))),$$

or, taking into account that $\pi$ is an automorphism and that $|\pi S| = |S|$, we have

$$\Phi_{\pi i}(v) = \sum_{S \subset I} \frac{(n - |S|)!(|S| - 1)!}{n!}(v(S) - v(S \setminus i)).$$

which in view of (17.9) shows that the *axiom of symmetry* is also valid.

3 *Aggregation*. This axiom follows directly from the fact that $\Phi$ is *linearly* represented in terms of the values of the characteristic function.

To summarize, we have shown that Shapley's vector does exist, is unique for any given characteristic function and its components are expressed by formula (17.9).

*4.17.10* **Theorem.** *The Shapley vectors of strategically equivalent characteristic functions correspond to each other.*

Proof. Let the characteristic functions $v$ and $v'$ be strategically equivalent, i.e. let for any coalition $K \subset I$ and for some $k > 0$ and $c_i$ $(i \in I)$,

$$v'(K) = kv(K) + \sum_{j \in K} c_j. \tag{17.13}$$

Setting $R = i \in I$ in the definition of the simplest characteristic function (cf. 4.17.1), we obtain

$$v_i(K) = \begin{cases} 1 & \text{if } i \in K \\ 0 & \text{otherwise} \end{cases}$$

for any coalition $K \subset I$.

Thus the identity (17.13) can be rewritten as

$$v'(K) = kv(K) + \sum_{j \in I} c_j v_j(K).$$

The linearity of the Shapley vector implies

$$\Phi_i(v') = k\Phi_i(v) + \sum_{j \in I} c_i \Phi_i(v_j) = k\Phi_i(v) + c_i,$$

which proves the correspondence between the vectors. □

*4.17.11.* We shall now describe an alternative representation for the components of the Shapley vector. Consider an arbitrary permuation $\pi$ on the set of players. Denote by $K_i(\pi)$ the set of all players preceding player $i$

in the permutation $\pi$. The difference

$$v(K_i(\pi)\cup i)-v(K_i(\pi))=\Delta(v,i,\pi) \tag{17.14}$$

is the increment in the value of the characteristic function $v$ "contributed" by player $i$ in the permutation $\pi$.

**Theorem.** *If all the permutations $\pi$ are equiprobable, then*

$$\underset{\pi}{M}\,\Delta(v,i,\pi)=\frac{1}{n!}\sum_{n}\Delta(v,i,\pi)=\Phi_i(v). \tag{17.15}$$

($\underset{\pi}{M}$ denotes the expectation with respect to $\{\pi\}$.)

PROOF. The number of permutations for which the only players preceding player $i$ are the players belonging to coalition $S$ is equal to

$$|S|!(n-1-|S|)!$$

and the probability of such a permutation is

$$\frac{|S|!(n-1-|S|)!}{n!}.$$

Consequently, in view of (17.14),

$$\underset{\pi}{M}\,\Delta(v,i,\pi)=\sum_{S\subset I\setminus i}\frac{|S|!(n-1-|S|)!}{n!}(v(S\cup i)-v(S)),$$

or changing the notation for variables in summation and denoting the coalition $S\cup i$ by $S$, we obtain

$$\underset{\pi}{M}\,\Delta(v,i,\pi)=\sum_{S\subset I}\frac{(|S|-1)!(n-|S|)!}{n!}(v(S)-v(S\setminus i)),$$

which proves the assertion. □

*4.17.12.* The expression for the Shapley vector, in terms of the difference $\Delta(v,i,\pi)$ derived in the preceding item, admits the following intuitive interpretation: Assume that the players arrive one after another in random sequence to a certain location and each receives a payoff equal to the amount by which his addition to the coalition of players assembled prior to his arrival increases the value of the characteristic function for the coalition. (This increment is referred to as an *admission value* for player $i$.) Then the expected value of the admission value for player $i$, when all the permutations of arrival sequences are equiprobable, is the $i$th component of the Shapley vector.

## 4.18 Examples of Shapley vectors

*4.18.1 A game with a foreman.* This is an $n$-person game and one of the players is called a *foreman.* Coalition $K$ wins one unit if it contains a foreman and at least one other player, or if its consists of $n-1$ "ordinary" players. If the foreman is designated as the $n$th player, then the characteristic function of the game can be written as follows:

$$v(K)=\begin{cases} 1, & \text{if } K\supset\{i,n\}, i\neq n, \\ 1, & \text{if } K\supset\{1,\ldots,n-1\}, \\ 0 & \text{otherwise} \end{cases}$$

Clearly, given any permutation of players, the foreman increases the payoff for the coalition by one unit provided he or she is neither the first nor the last (the $n$th) player in this permutation. The probability of this event is equal to $(n-2)/n$. In other cases, the foreman does not increase the payoff for a coalition. Consequently, the component of the Shapley vector for the foreman is

$$\Phi_{\text{for}}(v)=(n-2)/n.$$

Since the game is in the 0-1 reduced form, we have

$$\sum_{i=1}^{n-1}\Phi_i(v)=1-\Phi_n(v)=2/n.$$

It is, however, self-evident that all the ordinary players are equivalent. Therefore, by symmetry we have

$$\Phi_i(v)=\frac{2}{n(n-1)}, \qquad i=1,\ldots,n-1.$$

We thus observe that the "monopolistic" advantage of the foreman yields for him or her the payoff that is $(n-1)(n-2)/2$ times larger than the payoff for an "ordinary" player.

*4.18.2 "Landowner and farmhands."* Assume that there are $n-1$ farmhands (players $i=1,\ldots,n-1$) and a landowner (player $n$). The landowner, by hiring $k$ farmhands, assures the payoff $f(k)$, while the farmhands on their own receive no payoff. These rules can be described by the following characteristic function:

$$v(K)=\begin{cases} f(|K|-1), & \text{if } n\in K, \\ 0 & \text{otherwise.} \end{cases}$$

It is easy to verify that if all the permutations of players are equally likely, then the landowner will appear in a given permutation with probability $1/n$ at the $k$th place ($k=1,\ldots,n$) and thus he or she will increase the value of the characteristic function by the amount $f(k-1)$. Therefore,

$$\Phi_n(v)=\frac{1}{n}\sum_{k=1}^{n-1} f(k).$$

In view of the axioms of effectiveness and the symmetry of all the farmhands, the remaining components of the Shapley vector are given by

$$\Phi_i(v)=\frac{1}{n-1}\left(f(n-1)-\frac{1}{n}\sum_{k=1}^{n-1} f(k)\right), \qquad i=1,\ldots,n.$$

*4.18.3 A one-product balanced market.* Consider a market that involves a set of sellers $P$ and a set of buyers $Q$. Each seller $k\in P$ possesses the amount $x_k$ of a certain item, while the buyer $l\in Q$ demands the amount $y_l$ of this item. The assumption that the market is balanced implies the equality

$$\sum_{k\in P} x_k=\sum_{l\in Q} y_l.$$

Such a market can be described by a cooperative game with the set of players $I=P\cup Q$ (we shall assume as usual that $|I|=n$) and the characteristic function

$$v(K)=\min\left\{\sum_{k\in P\cap K} x_k, \sum_{l\in Q\cap K} y_l\right\}. \tag{18.1}$$

The value of this function is the "size" of the deals that can possibly be concluded between the sellers and buyers appearing in a coalition $K$.

We now determine the Shapley vector for this cooperative game. Consider an arbitrary permutation of players $\pi$ and the corresponding "opposite" permutation $\pi^*$ in which the players appear in the reverse order. Choose an arbitrary seller $k$ and denote by $K_\pi(k)$ the set of all the members of the market that precede this seller in permutation $\pi$. Denote the difference

$$\sum_{l'\in K_\pi(k)\cap Q} y_{l'} - \sum_{k'\in K_\pi(k)\cap P} x_{k'}$$

by $d(k,\pi)$. From the definition of the characteristic function $v(k)$ found in (18.1), we have

$$v(K_\pi(k)\cup k)-v(K_\pi(k))=\begin{cases} 0, & \text{if } d(k,\pi)\leqslant 0,\\ d(k,\pi), & \text{if } 0\leqslant d(k,\pi)\leqslant x_k,\\ x_k, & \text{if } x_k\leqslant d(k,\pi).\end{cases}$$

It is easy to verify that

$$d(k,\pi)+d(k,\pi^*)=x_k. \tag{18.2}$$

Therefore,

$$v(K_\pi(k)\cup k)-v(K_\pi(k))+v(K_{\pi^*}(k)\cup k)-v(K_{\pi^*}(k))=x_k.$$

This means that by combining the summands in the sum (17.15) in such a manner that each permutation is combined with its opposite, we shall obtain $n!/2$ pairs of terms and the sum of terms in each pair equals $x_k$ in this case. Consequently, for each seller $k\in P$ we have

$$\Phi_k(v)=\frac{1}{n!}\cdot\frac{n!}{2}x_k=\frac{x_k}{2},$$

and analogously for each buyer $l\in Q$

$$\Phi_l(v)=y_l/2.$$

Therefore, in the case of a balanced one-product market, the share of each of the participants in the distribution of the increments of utilities resulting from an exchange between sellers and buyers depends only on his or her capital (in the monetary or "natural" form) and does not depend on the distribution of the capital among the individual members of the market. This implies that the price in a balanced one-product market should be formed outside the market, i.e. in the realm of output and production, which is in agreement with the labor theory of value and the laws of cost analysis.

# Exercises

### Exercise for Section 1.6

State and prove the analogs of the assertion in Section 1.6.3 for the case of infima.

### Exercise for Section 1.10

Carry out in detail the transition to mixed strategies in the inequalities given in (10.3′).

### Exercises for Section 1.11

**1.** Prove the second assertion of the lemma in Section 1.11.1.

**2.** Prove the second assertion of the lemma in Section 1.11.3.

**3.** Prove the second part of the theorem in Section 1.11.4.

### Exercise for Section 1.12

Show that **0** belongs to an arbitrary convex cone.

### Exercise for Section 1.13

Show that vector $Y$ constructed in part b of the lemma belongs to $S_n$.

### Exercise for Section 1.15

Carry out in detail the argument for player II that will lead him or her to the minimax loss given by (15.2).

## Exercises for Section 1.16

1. Prove the second part of (16.1).
2. Prove the right-hand side of inequality (16.2).
3. Prove the validity of (16.3′) under the stated conditions.

## Exercises for Section 1.17

Solve the following $2\times 2$ matrix games:

(1) $\mathbf{A}=\begin{pmatrix} 1 & 0 \\ 0 & \frac{1}{2} \end{pmatrix}$

(2) $\mathbf{A}=\begin{pmatrix} 1 & -1 \\ -1 & 1 \end{pmatrix}$

(3) $\mathbf{A}=\begin{pmatrix} 0.8 & 1 \\ 1 & 0.6 \end{pmatrix}$

## Exercises for Section 1.18

1. Describe (graphically) the procedure for determining optimal strategies for player II when the abscissa of the "peak" is 1.
2. Solve the following $2\times 3$ matrix games:

$$\text{(a) } \mathbf{A}=\begin{pmatrix} 0 & \frac{5}{6} & \frac{1}{2} \\ 1 & \frac{1}{2} & \frac{3}{4} \end{pmatrix}; \qquad \text{(b) } \mathbf{A}=\begin{pmatrix} 3 & 4 & 12 \\ 8 & 4 & 2 \end{pmatrix}.$$

## Exercises for Section 1.19

Solve the following $3\times 2$ matrix games:

$$\text{(a) } \mathbf{A}=\begin{pmatrix} 1 & 4 \\ 2 & 3 \\ 3 & 0 \end{pmatrix}; \qquad \text{(b) } \mathbf{A}=\begin{pmatrix} 7 & 2 \\ 2 & 9 \\ 9 & 0 \end{pmatrix}.$$

## Exercises for Section 1.20

1. Complete the proof of the theorem in Section 1.20.1.
2. Complete the proof of the theorem in Section 1.20.5.
3. Complete the proof of the theorem in Section 1.20.8.
4. Explain the intuitive meanings of the theorem in Section 1.20.8 and its converse proved in Section 1.20.10.

## Exercises for Section 1.21

**1.** Define the strict dominance notion for pure strategies.

**2.** Prove the theorems stated in Sections 1.21.4 and 1.21.5 for the case of player II.

## Exercise for Section 1.22

Try to solve a diagonal game in which one of the entries on the diagonal is negative. (Start with $n=2$ and $n=3$.)

## Exercise for Section 1.23

Carry out the proof of the corresponding four properties for $T_2(\mathbf{A})$.

## Exercise for Section 1.24

Determine, using the method described in Section 1.24, the optimal strategies for the payoff matrix $\mathbf{A}=\begin{pmatrix} 0 & 2 & 0 \\ 1 & 1 & 2 \\ 2 & 0 & 0 \end{pmatrix}$.

## Exercises for Section 1.26

**1.** Given the following pair of dual linear programming problems:

$$\begin{array}{ll} 2x_1+x_2-2x_3 \leqslant 0 & \qquad 2y_1+y_2+y_3 \geqslant 5 \\ x_1+x_2-3x_3 \leqslant 0 & \qquad y_1+y_2+y_3 \geqslant 4 \\ x_1+x_2 \leqslant 1000 & \qquad -2y_1-3y_2 \geqslant -3 \\ \text{Maximize } 5x_1+4x_2-3x_3. & \qquad \text{Minimize } 0y_1+0y_2+100y_3. \end{array}$$

Determine the optimal solution of the corresponding game.

**2.** Solve the following $2\times 2$ games:

(a) $\mathbf{A}=\begin{pmatrix} -3 & 2 \\ 5 & -8 \end{pmatrix}$,

(b) $\mathbf{B}=\begin{pmatrix} 4 & 0 \\ 3 & -2 \end{pmatrix}$.

**3.** Solve the following $3\times 3$ game known as the *Scissors–Paper–Stone* game: Two players simultaneously name one of the three objects. If both name the same object, the game is a draw. Otherwise, the following nontransitive relation is valid: scissors cuts paper, stone breaks scissors, and paper covers stone.

The payoff matrix is:

| | | player II | | |
|---|---|---|---|---|
| | | Scissors | Paper | Stone |
| player I | Scissors | 0 | 1 | −1 |
| | Paper | −1 | 0 | 1 |
| | Stone | 1 | −1 | 0 |

**4.** Solve the following $5 \times 4$ game known as *Colonel Blotto's* problem: The Colonel has six companies and the enemy has five. They are engaged at two points, in company strength. The alternatives for the Colonel are to divide his forces in any one of these ways: 5:1, 4:2, 3:3, 2:4, 3:2, 2:3, or 1:4. We assume that equal forces result in a draw and the payoff is the number of companies defeated minus the number lost.

The corresponding payoff matrix is:

| | | player II | | | |
|---|---|---|---|---|---|
| | | 4:1 | 3:2 | 2:3 | 1:4 |
| | 5:1 | 4 | 2 | 1 | 0 |
| | 4:2 | 1 | 3 | 0 | −1 |
| player I | 3:3 | −2 | 2 | 2 | −2 |
| | 2:4 | −1 | 0 | 3 | 1 |
| | 1:5 | 0 | 1 | 2 | 4. |

Find the solution of the game. (*Hint:* Consider the subgame that was obtained by dropping the strategy 2:4 for player I.)

## Exercises for Section 2.2

**1.** Compute $\varepsilon$-optimal strategies for the players in the game described in Section 2.1.2 for $\varepsilon = \frac{1}{100}$ and find the corresponding "value" of the game.

**2.** Show that the minimax theorem does hold for the infinite two-player game defined by the payoff matrix $H(x,y) = \operatorname{sgn}(x-y)$, where $x = 1,2,\ldots$ and $y = 1,2,\ldots$.

**3.** Discuss the possible "optimal" strategies for the game given by $H(x,y) = x - y$, where $x = 1,2,\ldots$ and $y = 1,2,\ldots$. (Hint: Consider the strategy for player I given by

$$x_i = \begin{cases} 1/2i & \text{if } i = 2^k,\ k \text{ integer} \\ 0 & \text{otherwise.} \end{cases}$$

## Exercise for Section 2.3

Prove inequality (3.2) under the conditions stipulated in the first assertion of the theorem.

## Exercises for Section 2.4

**1.** Prove the second part of the lemma stated in Section 2.4.7.

**2.** Prove the second part of the lemma stated in Section 2.4.8.

**3.** Prove the second part of the lemma stated in Section 2.4.9.

## Exercises for Section 2.5

**1.** Prove the right-hand side of inequality (5.1).

**2.** Prove the second part of the theorem stated in Section 2.5.3.

**3.** Prove the second part of the theorem stated in Section 2.5.4.

**4.** Prove the second part of the theorem stated in Section 2.5.5.

## Exercise for Section 2.6

Verify that (6.1′) satisfies the triangular inequality.

## Exercises for Section 2.8

**1.** Show in detail that the inequality

$$H(X_\varepsilon, y_j) \leqslant H(X_\varepsilon, y) + \varepsilon$$

is valid for any $\varepsilon$ and any $y$ and an appropriately chosen $y_j$.

**2.** Show that the following $2\times2$ game on the *unit square* does not have a value:

$$H(x,y) = \begin{cases} -1 & x<y\neq1 \quad \text{and } x=1, y\neq1 \\ 0 & x=y \\ 1 & y<x\neq1 \quad \text{and } x\neq1, y=1. \end{cases}$$

**3.** Discuss the properties of the $2\times2$ antagonistic game on the unit square

$$H(x,y) = \begin{cases} 1 & \text{if } \; y\leqslant x<1 \quad \text{or} \quad x=y=1 \\ 0 & \text{otherwise.} \end{cases}$$

## Exercises for Section 2.9

**1.** Prove the second part of the lemma in Section 2.9.3.

**2.** Prove the basic theorem stated in Section 2.9.4 for player II.

**3.** Prove the second part of the theorem stated in Section 2.9.6.

## Exercises for Section 2.12

**1.** Prove the lemma stated in Section 2.12.3.

**2.** Verify the second case of the theorem in Section 2.12.5.

## Exercise for Section 2.14

Solve the game on the unit square given by the payoff function $H(x,y)=e^{-k(x-y)^2}$ Consider separately the case $0<k\leqslant2$ and $k>2$. (In particular, discuss case $k=3$.)

## Exercise for Section 3.4

Show in detail that the set of equilibrium situations $\{s^k\}$ in the Example is not a rectangular set and that $s^k$ are the only equilibrium situations in *pure* strategies.

## Exercise for Section 3.6

Carry out in detail the argument in three cases indicated in Section 3.6.6 leading to the determination of admissible situations for player II.

## Exercise for Section 3.7

Solve the following 2×2 bi-matrix games:

$$\mathbf{A}=\begin{pmatrix} -1 & 1 \\ 2 & -2 \end{pmatrix}, \qquad \mathbf{B}=\begin{pmatrix} -1 & 1 \\ -2 & 2 \end{pmatrix};$$

$$\mathbf{A}=\begin{pmatrix} 2 & -3 \\ -1 & 4 \end{pmatrix}, \qquad \mathbf{B}=\begin{pmatrix} 4 & -2 \\ -5 & 1 \end{pmatrix};$$

and

$$\mathbf{A}=\begin{pmatrix} 1 & -2 \\ 2 & -5 \end{pmatrix}, \qquad \mathbf{B}=\begin{pmatrix} 1 & 2 \\ -2 & -5 \end{pmatrix}.$$

The last game is called "chicken" (see, e.g., Rappaport [3E]).

## Exercise for Section 3.10

Verify in detail the validity of formulas (10.5) and (10.5′).

## Exercise for Section 3.11

Carry out in detail derivations leading to (11.2) by writing out explicitly the corresponding $H_i(\cdot, K^i)$ and $\sigma(K^i)$.

## Exercises for Section 3.12

1. Carry out in detail the calculations leading to (12.1) and (12.2).
2. Determine explicitly all the equilibrium situations for the game described in this section.

## Exercise for Section 4.2

Determine the characteristic function of the game discussed in Section 3.11 when the first two firms form a coalition and decide to advertise one product that yields a payoff of two units (daytime advertising) and a payoff of three units at night.

## Exercise for Section 4.4

Verify in detail the sufficiency part of the theorem in Section 4.4.4.

## Exercises for Section 4.8

1. Show that every essential game possesses a unique $a$-$b$ reduced form (provided $na \neq b$).

**2.** Show that every essential game possesses a unique $-1$-0 reduced form.

## Exercises for Section 4.9

**1.** Discuss in detail the geometric structure of the class of strategic equivalence for five-person constant sum cooperative games.

**2.** Discuss the class of strategic equivalence for four-person cooperative games not assuming constancy of the sum.

## Exercises for Section 4.10

Discuss the geometric structure of sets dominated by a given imputation $x$ if:

1. the first inequality is violated in (10.3).
2. the second inequality is violated in (10.3).
3. the first two inequalities are violated in (10.3).
4. the first and the third of the inequalities in (10.3) are violated.
5. Show that the three imputations: $x^1=(\frac{5}{8},\frac{1}{8},\frac{1}{8},\frac{1}{8})$, $x^2=(\frac{1}{2},\frac{3}{8},0,\frac{1}{8})$ and $x^3=(\frac{1}{4},\frac{1}{4},\frac{1}{4},\frac{1}{4})$ are nontransitively dominated in the game defined by $v(i)=0$, $i=1,2,3,4$, $v(2,3)=v(3,4)=v(2,4)=0$ while $v(K)=1$ for all the other coalitions $K\subset\{1,2,3,4\}$.

## Exercises for Section 4.11

**1.** Verify that the vector defined by (11.1′) is an imputation.

**2.** Verify that this vector dominates imputation $x$ through $K$.

## Exercises for Section 4.12

**1.** Classify in detail geometric shapes of the core for three-person (nonconstant sum) games according to the values taken on by $c_1=v(2,3)$, $c_2=v(1,3)$, and $c_3=v(1,2)$.

**2.** Consider the following three-person cooperative game:

$$v(\varnothing)=0.$$

$$v(1)=v(2)=v(3)=2.$$

$$v(23)=v(13)=v(12)=0.$$

$$v(123)=3.$$

(a) Is it a constant sum game?
(b) Is it a superadditive game?
(c) Show that the core of this game is an empty set.

**3.** Show that the game given by the characteristic function

$$v(\varnothing)=0 \quad v(1)=-1 \quad v(2)=-1 \quad v(3)=-1$$

$$v(23)=0 \quad v(12)=0 \quad v(13)=0$$

$$v(123)=3$$

has a nonempty core and determine it.

**4.** Determine the core of the game with $v(i)=0$ $(i=1,2,3)$; $v(123)=1$ and $v(23)=v(13)=v(12)=1$.

## Exercises for Section 4.14

**1.** Construct explicitly a vN–M solution in the case when the imputations of the solution lie on a single line and player one is discriminated.

**2.** Determine the sets of vN–M solutions of the following three-person game defined by the characteristic function:

$$v(1)=-2, \quad v(2)=v(3)=-4.$$
$$v(23)=2, \quad v(13)=v(12)=4.$$
$$v(123)=v(\varnothing)=0.$$

**3.** Consider the game

$$v(1)=v(2)=v(3)=0.$$
$$v(12)=v(13)=v(23)=1.$$
$$v(123)=1.$$

(cf. Exercise 4 for Section 4.12.)

(a) Show that every imputation of this game is in some vN–M solution.
(b) Show that the solutions of this game consist of

$$R_0=\left\{\left(0,\tfrac{1}{2},\tfrac{1}{2}\right),\left(\tfrac{1}{2},0,\tfrac{1}{2}\right),\left(\tfrac{1}{2},\tfrac{1}{2},0\right)\right\}$$

and, for all $i=1,2,3$ and all constants $c$ such that $0 \leqslant c < \frac{1}{2}$, of

$$R_i^{(c)} = \text{the set of all the imputations } (x_1,x_2,x_3) \quad \text{with } x_i=c.$$

(c) Describe this set geometrically.

## Exercises for Section 4.15

**1.** Consider a three-person constant sum cooperative game between three players labeled 1, 2, and 3. \$100.00 will be given to any coalition of two or three players if this particular coalition will agree on how to split the \$100.00 between the three players. If no such coalition (of at least two players) is formed, then each player receives nothing.
(a) Determine the "values" of various coalitions.
(b) Determine all the imputations belonging to the vN–M solutions

**2.** Consider the three-person cooperative game with

$$v(123)=v(12)=v(13)=1$$

and

$$v(23)=v(1)=v(2)=v(3)=v(\varnothing)=0.$$

(a) Determine the core for this game.
(b) Determine the vN–M solutions for this game.

## Exercise for Section 4.16

Show that the function $v'+v''$ defined by (16.3) is a characteristic function.

## Exercises for Section 4.17

**1.** Evaluate the Shapley vector for an inessential game.

**2.** Evaluate the Shapley vector for a two-person game with characteristic function $v$.

**3.** Evaluate the Shapley vector for an essential three-person constant sum game given in its 0-1 reduced form.

**4.** Evaluate the Shapley vector for a four-person game given by the characteristic function:

$$v(\varnothing)=0, \qquad v(i)=0 \quad (i=1,2,3,4)$$
$$v(23)=v(34)=v(24)=0, \qquad v(1234)=1,$$

and

$$v(ij)=v(ijk)=1$$

for all the other two- or three-person coalitions.

## Exercise for Section 4.18

Verify equation (18.2).

# Selected bibliography

## Elementary texts and treatises: (E)


1. Davis, M. D., 1970, *Game Theory, A Nontechnical Introduction*, New York, Basic Books
2. May, F. B., 1970, *Introduction to Games of Strategy*, Boston, Allyn and Bacon
3. Rappoport, A., 1966, *Two-person Game Theory, The Essential Ideas*. Ann Arbor, Michigan, University of Michigan Press
4. Rappoport, A., 1970, *N-person Game Theory, Concepts and Applications*, Ann Arbor, Michigan, University of Michigan Press
5. Shubik, M. (ed.), 1964, *Game Theory and Related Approaches to Social Behavior, Selections*, New York, J. Wiley
6. Vajda, S., 1967, *The Theory of Games and Linear Programming*, London, Methuen (Science Paperbacks)
7. Ventzel, E. S., 1961, *Lectures on Game Theory*, Delhi, Hindustan Publ. Corp. (a translation from Russian)
8. Vorob'ev, N. N., 1976, *Theory of Games*, Moscow, Znanie [Russian].
9. Williams, J. D., 1966, *The Compleat Strategist*. New York, McGraw-Hill (revised edition)


## Intermediate level texts (I)


1. Berge, C., 1957, *Théorie Generale des Jeux à n Personnes*, Paris, Gauthier-Villars.
2. Burger, E., 1963, *Introduction to the Theory of Games*, (translated from German by J. E. Freund), Englewood Cliffs, Prentice-Hall
3. Luce, R. D., Raiffa, H., 1957, *Games and Decisions, Introduction and Critical Survey*, New York, John Wiley
4. McKinsey, J. C. C., 1952, *Introduction to the Theory of Games*, New York, McGraw-Hill

5. Owen, G., 1968, *Game Theory*, Philadelphia, W. B. Saunders
6. Vorob'ev, N. N., 1975, *Entwicklung der Spieltheorie*, Berlin, Deutscher Verlag der Wissenschaften.

## Advanced and specialized texts (A)

1. Aumann, R. J., Shapley, L. S., 1974, *Values of Nonatomic Games*, Princeton, N.J., Princeton University Press
2. Karlin, S., 1959, *Mathematical Methods and Theory in Games, Programming, and Economics*, Reading, Mass., Addison Wesley
3. von Neumann, J., Morgenstern, O., 1953, *Theory of Games and Economic Behavior*, 3rd ed., Princeton, N.J., Princeton University Press
4. Parthasarathy, T., Raghavan, T. E. S., 1971, *Some Topics in Two-person Games*, New York, American Elsevier
5. Rosenmüller, J., 1971, *Kooperative Spiele und Märkte*, Berlin-Heidelberg-New York, Springer-Verlag.

## Basic survey papers (S)

1. Lucas, W. F., 1971, Some recent developments in $n$-person game theory, *SIAM Review* 13, 491–523
2. Vorob'ev, N. N., 1970, The present state of game theory, *Uspehi Mat. Nauk* 25, no. 2 (152), 81–140 [Russian], (English translation in *Russian Mathematical Surveys* 25, no. 2, 78–136.)
3. Vorob'ev, N. N. (ed.), 1976, *Game Theory, Annotated Index of Publications up to* 1968, Leningrad, Nauka [Russian].

# Index

## U



## V



## Z

## Applications of Mathematics



Volume 1
W. H. Fleming and R. W. Rishel
**Deterministic and Stochastic Optimal Control**
1975. ix, 222p. 4 illus. cloth

Volume 2
G. I. Marchuk
**Methods of Numerical Mathematics**
1975. xii, 316p. 10 illus. cloth

Volume 3
A. V. Balakrishnan
**Applied Functional Analysis**
1976. x, 309p. cloth

Volume 4
A. A. Borovkov
**Stochastic Processes in Queueing Theory**
1976. xi, 280p. 14 illus. cloth

Volume 5
R. S. Liptser and A. N. Shiryaev
**Statistics of Random Processes I**
**General Theory**
1977. approx. 300p. cloth

Volume 6
R. S. Liptser and A. N. Shiryaev
**Statistics of Random Processes II**
**Applications**
In preparation.

Springer-Verlag
New York Heidelberg Berlin